PLUMBING MATERIALS AND DRINKING WATER QUALITY

PLUMBING MATERIALS
AND
DRINKING WATER QUALITY

Thomas J. Sorg

U.S. Environmental Protection Agency
Drinking Water Research Division
Water Engineering Research Laboratory
Office of Research and Development
Cincinnati, Ohio

Frank A. Bell, Jr.

U.S. Environmental Protection Agency
Criteria and Standards Division
Office of Drinking Water
Washington, D.C.

NOYES PUBLICATIONS
Park Ridge, New Jersey, U.S.A.

Published in the United States of America by
Noyes Publications
Mill Road, Park Ridge, New Jersey 07656

10 9 8 7 6 5 4 3 2 1

Library of Congress Cataloging-in-Publication Data

Sorg, Thomas J.
    Plumbing materials and drinking water quality.

    (Pollution technology review, ISSN 0090-516X ;
no. 128)
    Bibliography: p.
    Includes index.
    1. Drinking water--Equipment and supplies.
2. Plumbing--Equipment and supplies. 3. Water
quality management. I. Bell, Frank A. II. Title.
III. Series.
TD353.S67   1986   363.6'1   85-25840
ISBN 0-8155-1066-7

Foreword

A review of drinking water problems related to plumbing materials and the identification of alternative solutions for dealing with these problems are presented in this book. It should be of interest to scientists, engineers, and technicians involved in water-related activities.

The information in the book is from *Plumbing Materials and Drinking Water Quality: Proceedings of a Seminar, Cincinnati, Ohio, May 16-17, 1984;* co-sponsored by the Office of Drinking Water, U.S. Environmental Protection Agency, and the Drinking Water Research Division, Water Engineering Research Laboratory, Office of Research and Development, U.S. Environmental Protection Agency; issued February 1985. Thomas J. Sorg, USEPA Cincinnati, and Frank A. Bell, Jr., USEPA Washington, were co-chairmen of the seminar, which was coordinated by Eastern Research Group, Inc., Cambridge, Massachusetts.

The table of contents is organized in such a way as to serve as a subject index and provides easy access to the information contained in the book.

Preface

The National Interim Primary Drinking Water Regulations define a Maximum Contaminant Level (MCL) as "the maximum permissible level of a contaminant in water which is delivered to the free-flowing outlet of the ultimate user of a public water system, except in the case of turbidity where the maximum permissible level is measured at the point of entry to the distribution system." With this definition, the Federal regulations clearly recognize that the quality of drinking water can be affected by its distribution lines and that it is the responsibility of the water purveyor to consider these problems in providing water to its customers.

Traditionally, water utilities have taken control measures to prevent or correct problems associated with water distribution mains. Studies and reports during the past several years have clearly illustrated that distribution problems are not limited only to distribution mains, but are also associated with laterals and service lines as well as the interior plumbing systems in dwellings and buildings. Some of these problems, such as lead leaching from lead service lines and plumbing, have been common knowledge for years. Other problems, such as the migration of gasoline and other petroleum distillates through plastic pipe, have been more recently identified.

On May 16–17, 1984, the Office of Drinking Water and the Drinking Water Research Division of the U.S. Environmental Protection Agency cosponsored a seminar to review drinking-water problems related to plumbing materials (including those used in service lines) and to identify solutions for dealing with these problems. In attendance were approximately 150 people representing government, manufacturers, trade associations, consultants, and public interest organizations. During the first day, speakers addressed three general topics: (1) the use, application, and availability of plumbing materials (metallic and plastic); (2) the impact of these materials on water quality; and (3) solutions to plumbing problems related to water quality.

On the second day, four concurrent panel sessions were held for all attendees to share their knowledge and to express their viewpoints on the information presented. These panel sessions proved to be very worthwhile. A wide variety of opinions and viewpoints were expressed on the four panel topics: (1) joining alternatives for copper pipe; (2) metal pipe and fitting alternatives for plumbing; (3) plastic pipe and fittings; and (4) regulatory and compliance aspects.

The first section of these Proceedings [Parts I–IV] includes the papers given by the speakers on plumbing material applications, problems, and solutions. Part IV of this section includes as a highlight a review of "European Developments in Use of Plumbing Materials," presented by Ivo Wagner of the Engler-Bunte Institute, Karlsruhe, West Germany. The second section [Part V] contains the reports of the panel sessions prepared by the panel chairmen. Although these reports vary somewhat in format, each contains a summary of the opinions, conclusions, and recommendations of the panel. The Proceedings conclude with a summary [Part VI] of the conclusions and recommendations of the panel meetings and a synopsis of the seminar presentations. This seminar was timely and successfully presented many views on the complex issue of plumbing materials and water quality.

Frank A. Bell, Jr., Co-chairman
Criteria and Standards Division
Office of Drinking Water

Thomas J. Sorg, Co-chairman
Drinking Water Research Division
Water Engineering Research
    Laboratory
Office of Research and Development

Contents and Subject Index

PART II
IMPACT OF METALLIC PLUMBING MATERIALS ON WATER QUALITY

IMPACT OF LEAD PIPING AND FITTINGS ON DRINKING WATER QUALITY

Peter C. Karalekas, Jr.

Chester H. Neff

Norman E. Murrell

PART III
IMPACT OF PLASTIC PIPE AND FITTINGS ON WATER QUALITY

PART IV
SOLUTIONS TO PLUMBING-RELATED WATER QUALITY PROBLEMS

**PART VI
SUMMATION**

Part I

Plumbing Materials: Extent of Use, Application, and Availability

METAL PIPING AND JOINING MATERIALS AND FITTINGS

Paul A. Anderson

Vice President
Copper Development Association, Inc.
Greenwich Office Park No. 2
Greenwich, Connecticut 06836

Copper is the oldest engineering metal, and has been in continuous use for plumbing since the time of the Pharaohs. Copper sheet rolled into tubular form to conduct water has been found in excavated Egyptian tombs. Over the centuries water has been conducted to the consumer through stone aqueducts, hollow wooden logs, iron and steel pipe, and today's highly engineered copper tubular products.

The Copper Development Association analyzes the 25 largest application areas for copper and its alloys. Arranged in order of rank, based on 1983 data, plumbing and heating is the most important use for copper, followed by building wiring. These two major building construction applications accounted for 27 percent of U.S. copper and copper alloy shipments last year. This is the first time plumbing and heating has led the list.

Each year in the United States, well over half a billion linear feet of copper water tube are installed in water service and distribution systems. This is equivalent to about 125,000 miles of hot and cold water systems. Since 1946, when such statistics began to be gathered in the United States, 17 billion feet of copper tube have been installed in water service and distribution systems for buildings. That is more than 3 million miles.

But what do these copper shipments mean? The market can be measured by usage intensity. Is copper usage in plumbing growing or declining compared to building construction in general? To find out, we need to compare copper plumbing tube usage to building construction activity.

Housing starts seem to be the best measure of total building activity. Plumbing is predominantly a residential market in the United States, and using square feet of total construction makes the index too diffuse. The two curves in Figure 1 compare housing starts and copper plumbing tube shipments over the last 12 years on an index basis, with 1972 set equal to 100. There is no doubt that copper plumbing tube consumption

140
Index
120
100
80
60
U.S. Plumbing & Heating Market
COPPER PLUMBING TUBE VS. HOUSING STARTS
100
Copper
Housing Starts
72
75
80
83

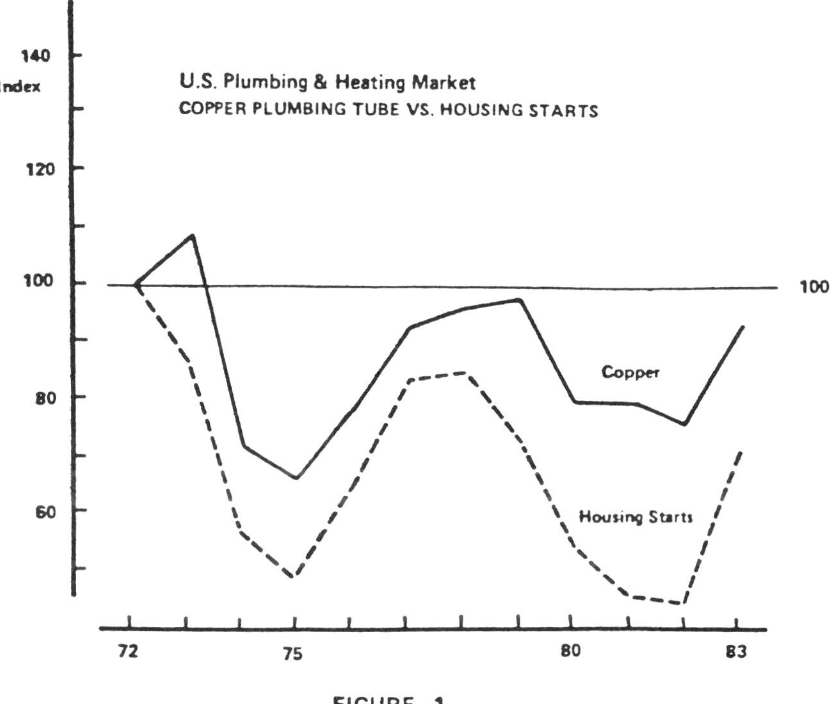

FIGURE 1

Copper Usage Intensity
U.S. PLUMBING & HEATING MARKET
180
Index
160
140
120
100
72
75
80
83
100

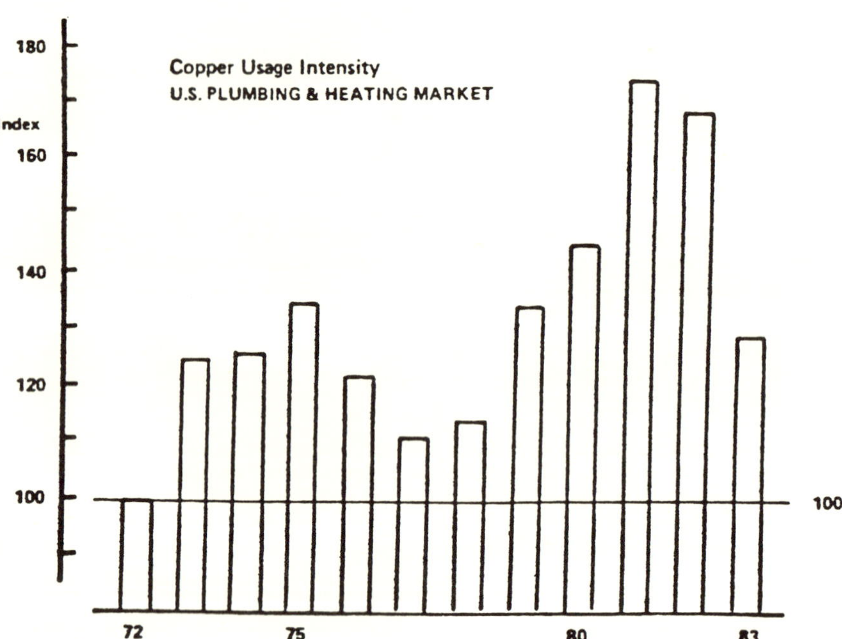

FIGURE 2

follows housing starts. To convert these data to usage intensity, we can divide the index for copper shipments by the index for housing starts.

This gives an intensity index -- usage per housing start, in effect (Figure 2). The result highlights something that was difficult to appreciate when it was happening in 1974 and 75 and in 1980, 1981, and 1982. It shows that plumbing and heating has been a very healthy usage for copper over the last 10 years, outperforming housing starts. How can this be explained?

The major factors include the introduction of copper fire sprinkler and solar heating systems, which brought copper technology into new applications in the market in the mid-1970's. There has been increased usage of large diameter copper tube starting in 1981. Large diameter tube has, in the past few years, penetrated plumbing applications held by steel, largely in commercial and industrial buildings. Finally, there seems to be a trend, as in the 1960s, for more plumbing per dwelling unit.

Through the late 1960s and the 1970s, increasing copper usage intensity in the plumbing and heating market trended downward. This measure reached a low point in 1977. It reflected the well-known shift to thinner wall tube in domestic plumbing systems. This was plumbing progress that improved copper's usage efficiency -- more plumbing per pound of copper. Since 1977, however, the downward trend has been overpowered by increasing use of larger diameter and thus heavier weight copper tube. Fire sprinklers and solar heating use larger diameters than domestic water systems. Large diameter tube usage is increasing in commercial and industrial water supply systems.

To appreciate the relationship between the end-use market and the reliability of the plumbing tube product, understanding of the manufacture and quality assurance operations is needed. Copper metal is derived from two sources, the mine and recycled scrap, each accounting for about 50 percent of total U.S. production.

In fact, copper is the most recycleable of all the engineering metals, with the use of recycled scrap averaging between 44 and 54 percent of total production over the past decade; the figure was 53 percent last year. The reason: recycled copper is identical chemically and metallurgically with new refined copper.

Copper cathode from the refinery is blended with recycled No. 1 scrap from either wire and cable mills or brass mills, then remelted to produce the copper billet. These billets are held in the cast shop until the production quality control laboratory has completed its quality analysis and releases the billet for conversion into tubular products in the mill.

Massive equipment is involved in the conversion of the billet into tubular products. Following inspection, each billet is heated to the proper temperature prior to insertion into the extrusion press for conversion into a tube shell. The ram is inserted into the billet cavity

and pressure applied to the heated billet, moving it to the die surface.
The billet is then pierced prior to operation of the main ram. The main
ram then pushes the hot billet through the die with the center punch of the
ram acting as the internal core around which the tube forms as it is
extruded through the die. Surface defects are prevented from being
extruded into the tube shell by proper ram and extrusion chamber clearance.

The tube shell is further reduced to final size through a series of
diameter/wall thickness reductions on bull blocks or draw benches. Drawing
lubricants used by seamless tube manufacturers are lead-free,
water-soluble, lightweight mineral oils. The nature of tube size reduction
is such that the internal surface is wiped clean of any residue by a
floating plug and tool steel die.

These tubular products are produced either as copper or copper alloys
by the seamless or welded manufacturing methods. A total of 12 ASTM
specifications have been developed to cover these commercial products.
Ninety-eight percent of water tube is manufactured according to ASTM
Standard Specification B88.

ASTM B88 describes three types of copper tube: Types K, L and M.
Each of these tube types is a series of standard diameter sizes with
specified wall thicknesses. Type K is the thickest wall, Type M the
thinnest, L intermediate. A lightweight distribution tube is also
available. About three quarters of all copper water tube installed is
either half-inch or three-quarter-inch in nominal size.

Prior to packaging and shipment, all tube must undergo product
inspection to ensure that the finished tube meets quality assurance
requirements. This is accomplished by the eddy-current, nondestructive
inspection operation, which is not subject to operator fatigue, so exacting
acceptance criteria are continually maintained on the production line.
Immediately after the eddy-current inspection station, the name of the
manufacturer and tube type are marked on the tube surface by inking and
incise impressions. This identifies the tube as to source, type, and
specification and assists the plumbing inspector in verifying that it meets
code requirements.

Solder joint plumbing fittings are produced according to the same
exacting manufacturing standards as tube (ANSI B16.18 or B16.22). They are
made in a wide variety of shapes and sizes, including couplings, T's,
elbows, and reducing and special fittings. All quality assurance
operations applied to tube apply here as well.

Up to now, my remarks have been directed to the role of copper tube in
use, its production and quality assurance. Let me now comment on why
copper has achieved such a remarkable level of acceptance by the building
community.

Copper is the dominant material used in potable water systems. Its
combination of corrosion resistance in hot and cold systems and ease of

installation is unmatched by any other plumbing material in use today.
Service experience in general indicates that copper plumbing tube will
outlast the building in which it is installed. Further, the combination of
nontoxicity and nonflammability properties makes it the leading choice.

As with all materials, copper can be attacked by some highly
aggressive environments, and problems arise occasionally with certain well
waters, with very soft waters high in carbon dioxide, and in systems where
the water flow velocity in excess of about 5 fps causes erosion.

In a modern waterworks, the chemist in charge not only prescribes
treatments to make the water conform to health regulations, such as to free
it of pathogenic organisms, but for well waters he must also make
arrangements, if necessary, for chemical treatments designed to reduce the
corrosiveness of the water. The water purveyor may not fully recognize
this responsibility or the economic advantages of providing a water
suitably adjusted chemically to reduce its corrosiveness. All of this is
directly relevant to one of the topics of this conference: lead pickup
from soldered plumbing systems.

Potable waters are developed from a variety of sources, including
lakes, rivers, streams, and wells. Reservoirs are used to store water
where there is seasonal or annual variation. Surface waters tend to be
saturated with air. Waters taken directly from wells may be low in oxygen
but may contain much higher levels of carbon dioxide and minerals than
surface supplies. Unlike surface waters, well waters are usually free of
algae and organic matter.

Waters are routinely treated for biological purity, but in addition,
each water source has to be examined to determine whether some form of
correction or treatment is necessary to control its corrosiveness. Often,
all that is required is pH adjustment. The form of corrective treatment
may be to control scaling or to reduce corrosion. Treatments with other
objectives must be evaluated for the possible effect on the corrosiveness
of the water. For example, water must meet the new EPA limit of one
turbidity unit. Flocculation, the treatment to clarify turbid water, often
uses a coagulant such as alum to remove suspended matter. Water treatment
can also be used to protect consumers from lead pickup by aggressive acid
waters.

Suitable treatments are routine in modern waterworks. Where the water
purveyor continuously controls the character of the water fed to the mains,
he not only ensures optimum corrosion performance of his own supply system
but also greatly reduces costly customer-related water problems.
Unfortunately, smaller water purveyors in many cases do not provide
corrective treatments, either from lack of understanding of the need or
from lack of the proper facilities.

Corrosion and metal pickup, including lead pickup from poorly made
solder joints, can be prevented in domestic plumbing systems by refraining
from the use of aggressive waters or with effective water treatment of

either surface or well waters. This has been the case for over 50 years of highly serviceable use of copper tube and fittings for water distribution systems. Examples of successful treatments that have been introduced include simple aeration, the addition of caustic soda, soda ash or lime. Neutralizing filters have also been used in private dwellings. Up to now, these treatments have been aimed primarily at protecting consumers against pitting and general corrosion, but the same principles apply for protection against lead pickup by acid soft waters from poorly made joints. Consider a few case histories that demonstrate the effectiveness of this approach to consumer protection.

At one location in California, simple aeration of the water reduced the free carbon dioxide content from over 20 mg/1 to about 10 mg/1 and was adequate to alleviate a corrosion problem in the system served. This was done simply by spraying the well water into the water supplier's large storage tank via a large-diameter perforated pipe at the top of the tank reservoir.

In another area treatment with caustic soda (NaOH) was effective in neutralizing the free carbon dioxide in a pitting water serving a particular housing development, causing the pitting attack to cease. When treatment was stopped, the pitting resumed. When a nonpitting water was finally introduced into the system, pitting stopped altogether. In this case, the treatment consisted of the addition of caustic soda to raise the pH and eliminate the free carbon dioxide.

Another example of the effectiveness of water treatment is in the community of Fort Shawnee, Ohio. Here, an isolated housing development of about 100 residences and condominiums is served by a single well-water source. About five years ago, an epidemic of cold water pitting in these homes was brought to our attention. Of the 100 or so dwellings, 16 suddenly experienced 28 pitting perforations in their cold water distribution systems.

Equipment was installed in the waterworks to inject sodium carbonate $(Na_2CO_3 \quad H_2O)$ into the water to raise the pH and neutralize the free carbon dioxide. Since the treatment began there has not been a single additional corrosion failure within the buildings served by the treated water.

In Suffolk County, New York, a water supply system with about 150 well fields (two to three wells per field) has for many years treated the combined output of more than 350 wells by injecting a slurry of lime (CaO) into the supply. The remaining wells are treated with caustic soda. When properly treated, this raises the pH to about 7.5, reducing the CO_2 to less than 5 mg/1, and eliminates the copper water tube corrosion problems on the system almost completely.

A housing development in San Bernardino County, California, started to experience corrosion approximately 1.5 years after the residences were initially occupied. Because of the reluctance of the water utility to

introduce the needed treatment, neutralizing filters were introduced by the
developer at each residence. These units contained a bed of dolomite
($CaCO_3$ MgO) which functions by reacting with and removing the free
carbon dioxide from the water with concurrent increase of pH of the water.

The excess hardness (Ca and Mg) is then reduced by passing the
first-stage treated water through a second tank containing an ion exchange
resin.

To summarize, copper tube and fittings have been the dominant accepted
plumbing products in the United States for several decades, providing
trouble-free service to residential and commercial users.

It would be foolhardy to claim that corrosion has not occurred in
copper plumbing systems, but our studies show this situation to be highly
localized and preventable. Other than for faulty design and poor
workmanship, corrosion and metal pickup are invariably associated with
aggressive water compositions. Successful treatments utilized by local
water utilities have corrected the water chemistry problems. This has
resulted in trouble-free plumbing systems for the consumer and high quality
drinking water. The same approach can protect consumers from lead pickup
from poorly made solder joints in those few — really extremely few and
isolated — potable water systems that are aggressive as a result of
acidity and lack of hardness.

PLASTIC PIPING AND JOINING MATERIALS

Stanley A. Mruk

Technical Director
Plastics Pipe Institute
A Division of The Society of the Plastics Industry, Inc.
355 Lexington Avenue, 6th Floor
New York, New York 10017

INTRODUCTION

First use of thermoplastics piping for conveying potable water dates back to the early 1950s. Since then, thermoplastics piping has grown to become a most widely used material for a broad range of applications including water distribution, sewerage, drainage, plumbing, gas distribution, industrial uses, and power and communications ducting. Experience has established that thermoplastics piping very often provides the most economical and satisfactory long-term solution.

The broad acceptance of thermoplastics piping not only evidences the inherent capabilities and versatility of these materials, but it also reflects the attention that has been given, from the start of the plastics industry, to the development of a standardization system that suitably addresses all the major performance requirements, including preservation of water quality. The key element of the standardization system covering thermoplastics piping for potable water applications is that it scrutinizes each and every commercial formulation for two most important characteristics: no adverse effect to water quality, and long-term strength under end-use conditions.

AVAILABLE MATERIALS FOR PLUMBING APPLICATIONS

The major thermoplastic materials used for potable water transport and plumbing applications include polyvinyl chloride (PVC), chlorinated polyvinyl chloride (CPVC), polyethylene (PE), polybutylene (PB), and acrylonitrile-butadiene-styrene (ABS). As indicated by Table 1, all of these materials are used to manufacture pressure piping. However, some materials and products are used preferentially over others depending on the application, and Table 1 also shows current market preferences.

TABLE 1

THERMOPLASTIC PIPING PRODUCTS
FOR POTABLE WATER APPLICATIONS

KEY: X – Most Frequently Used
        O – Less Frequently Used

MATERIAL	PRODUCT STANDARD	TITLE (ABBREVIATED) OF STANDARD	SIZE RANGE (INCHES)	MAINS		SERVICES	YARD PIPING	HOT/COLD DISTRIBUTING
				RURAL	MUNICIPAL			
PVC	AWWA C 900	PVC Pressure Pipe for Water	4 to 12	O	X			
	ASTM D 2241	PVC Plastic Pipe SDR/PR	1/2 to 24	X	O	O	X	
	ASTM D 2466	PVC Fittings, Sch. 40	1/2 to 8	X	O	O	X	
	ASTM D 2672	PVC Pipe, Belled End	1/2 to 8	X	O	O	X	
	ASTM D 1785	PVC Plastic Pipe, Sch. 40-80	1/2 to 12	O	O	O	O	
	ASTM D 2740	PVC Plastic Tubing	1/2 to 1-1/4			O		
PE	AWWA D 901	PE Pipe & Tubing for Services	1/2 to 3			X	O	
	ASTM D 2239	PE Plastic Pipe, SDR/PR	1/2 to 6	O	O	X	X	
	ASTM D 2609	Plastic Insert Fittings	1/2 to 4	O		X	X	
	ASTM D 2737	PE Plastic Tubing	1/2 to 2			X	O	
	ASTM D 2683	PE Fittings, Socket Type	1/2 to 4			O	O	
	ASTM D 3281	PE Fittings, Butt Type	1/2 to 10	O	O	O	O	
	ASTM D 3035	PE Pipe, SDR/PR, OD Controlled	1/2 to 6	O	O	O	O	
	ASTM F 714	PE Pipe, SDR/PR, Large Diam.	3 to 63	O	O			
PB	AWWA C 902	PB Pipe & Tubing for Services	1/2 to 3			X	O	
	ASTM D 2662	PB Pipe, SDR/PR	1/2 to 6			X	O	
	ASTM D 2666	PB Tubing	1/2 to 2			X	O	
	ASTM D 3000	PB Pipe SDR/PR, OD Controlled	1/2 to 6			O	O	
	ASTM D 3309	PB Hot/Cold Water Systems	1/2 to 2				X	
CPVC	ASTM D 2846	CPVC Hot/Cold Water Systems	1/2 to 2					X
	ASTM F 441	CPVC Pipe, Sch. 40-80	1/2 to 12					O
	ASTM F 442	CPVC Pipe, SDR/PR	1/2 to 12					O
	ASTM F 438	CPVC Fittings, Sch. 40	1/2 to 6					O
ABS	ASTM D 2282	ABS Pipe, SDR/PR	1/2 to 12	O		O		
	ASTM D 468	ABS Fittings, Sch. 40	1/2 to 8	O		O		

The major products used for the various non-pressure plumbing applications are presented in Table 2. Pressure and non-pressure piping made from ABS, PVC, and CPVC can be jointed by solvent cementing. Table 3 identifies the standard cements. The polyolefins, PE and PB, are most often joined by mechanical systems, primarily insert, compression, and flare. They can also be joined by heat fusion, but this technique is usually used for larger pipe and by major users such as gas distribution companies.

A very popular technique for joining buried PVC pipe is the elastomeric gasket. Almost all PVC American Water Works Association (AWWA) C-900 water distribution pipe as well as American Society for Testing and Materials (ASTM) D 3034 sewer pipe is manufactured with integral bells with gaskets, into which the spigot end may be pushed for easy joining in the field. The various available joining methods are summarized in Table 4.

Estimates of the approximate market share held by the various thermoplastic materials in water distribution and plumbing applications are given in Table 5. These are very rough estimates because accurate statistics on actual shipments to different market sectors presently are not collected by government or industry. The estimates presented in Table 5 have been developed by consolidating the best judgment of a number of knowledgeable people in the industry.

ATTRIBUTES OF PLASTIC PIPING

When evaluating the capacities and limitations of any piping material, it is important to examine its performance in an operational system rather than to compare certain individual properties with those offered by alternate materials. Individual properties of materials, such as shown in Table 6, are measured under conditions that do not sufficiently approximate those encountered in actual use. Furthermore, each material can have its own particular problems, all of which need to be recognized and resolved if the material is to be used successfully in a working system. With traditional materials, the resolution of many of these problems has come about very gradually through years of experience. With the newer thermoplastics materials, the evolution to mature engineering materials has been accelerated through an emphasis on the technological assessment of potential performance limitations, particularly long-term durability. That this evolution is proceeding most successfully is indicated by the quite successful application history and the excellent acceptance of plastics for a broad range of applications.

ADVANTAGES

The principal advantages of thermoplastics piping are as follows:

Immunity to "corrosion". Plastics do not rust, pit, or corrode in the sense that metals do. Being nonconductors, they are not susceptible to galvanic or electrochemical effects and, therefore, are unaffected by

TABLE 2

THERMOPLASTIC PIPING PRODUCTS
FOR NON-PRESSURE PLUMBING APPLICATIONS

KEY: X - Most Frequently Used
      O - Less Frequently Used

MATERIAL	PRODUCT STANDARD	TITLE (ABBREVIATED) OF STANDARD	SIZE RANGE (INCHES)	BUILDING DRAINAGE & VENT		BLDG. SEWER	BLDG. STORM SEWER	SUBSOIL DRAIN
				ABOVE GROUND	BELOW GROUND			
PVC	ASTM D 2665	PVC DWV Pipe & Fittings	1-1/4 to 6	X	X	X	X	O
	ASTM D 2949	3-In. PVC Thin Wall DWV Pipe	3	X	X	O		
	ASTM D 3311	DWV Fitting Patterns	1-1/4 to 6	X	X			
	ASTM D 2729	PVC Drain Pipe & Fittings	2 to 6			O	O	X
	ASTM D 3033	PVC Sewer Pipe & Fittings, PSP	4 to 15			O	O	O
	ASTM D 3034	PVC Sewer Pipe & Fittings, PSM	4 to 15			X	X	O
ABS	ASTM D 2661	ABS DWV Pipe & Fittings	1-1/4 to 6	X	X	X	X	O
	ASTM F 628	ABS Foam Core DWV	1-1/4 to 6	X	X	X	X	O
	ASTM D 2751	ABS Sewer Pipe & Fittings	1 to 6			X	O	O
PE	ASTM F 405	PE Corrugated Tubing & Fittings	3 to 8					X

TABLE 3

SOLVENT CEMENTS FOR THERMOPLASTIC PIPING

MATERIAL	PRODUCT STANDARD	TITLE (ABBREVIATED) OF STANDARD
ABS	ASTM D 2235	Solvent Cements for ABS Piping
PVC	ASTM D 2564	Solvents for PVC
ABS/PVC	ASTM D 3138	Solvent Cements for ABS/PVC Transitions
CPVC	ASTM F 493	Solvent Cements for CPVC Piping

TABLE 4

JOINING METHODS

METHOD	THERMOPLASTIC PIPE				
	ABS	PVC	CPVC	PE	PB
Solvent Cementing	O	O	O	-	-
Heat Fusion	-	-	-	O	O
Threading (Sch. 80 only)	O	O	O	O	-
Insert	-	-	-	O	O
Mechanical Compression	O	O	O	O	O
Flaring	-	-	-	O	O
Elastomeric Seal	O	O	O	O	O

TABLE 5

ESTIMATED MARKET SHARES
OF THERMOPLASTIC PIPING IN WATER
DISTRIBUTION AND PLUMBING APPLICATIONS

PRESSURE PIPE

| | MAINS | | | YARD | HOT/COLD |
MATERIAL	RURAL	MUNICIPAL	SERVICES	PIPING	DISTRIBUTING
PVC	80	25	10	20	
PE	1		30	60	
PB			10		6
CPVC					9
ABS					
Others	19	75	50	20	85
	100	100	100	100	100

NON-PRESSURE PIPE
(FOR HOME CONSTRUCTION ONLY)*

| | BUILDING DRAINAGE & VENT | | | BLDG. | |
MATERIAL	ABOVE GROUND	BELOW GROUND	BLDG. SEWER	STORM SEWER	DRAIN
PVC	60	60	70	70	40
ABS	20	20	10	10	
PE (Corrugated)					60
Others	20	20	20	20	
	100	100	100	100	100

*Market shares for plastics are substantially less for other types of construction.

TABLE 6

TYPICAL PHYSICAL PROPERTIES OF
MAJOR THERMOPLASTIC PIPING MATERIALS

PROPERTY AT 75° F	ASTM TEST NO.	ABS	PVC	CPVC	PE	PB
Specific Gravity	D-792	1.08	1.40	1.54	0.95	0.92
Tensile Strength psi (10^3)	D-638	7.0	8.0	8.0	3.2	4.2
Tensile Modulus psi (10^5)	D-638	3.4	4.1	4.2	1.3	0.55
Impact Strength, Izod ft-lb/inch notch	D-256	4	1	1.5	∣10	∣10
Coeff. of Linear Expansion in/in-°F (10^5)	D-696	6.0	3.0	3.5	9.0	7.2
Thermal Conductivity Btu-in/hr-ft-°F	D-177	1.35	1.1	1.0	3.2	1.5
Specific Heat Btu/lb-°F	–	0.34	0.25	0.20	0.55	0.45
Approx. Operating Limit[1]						
°F, nonpressure	–	180	150	210	160	210
°F, pressure	–	160	130	180	140	180

[1]Exact operating limit will vary for each particular commercial plastic material. Effect of environment should be considered.

acids, bases, salt, "corrosive" waters, or "hot" soils. They can be, however, susceptible to other forms of degradation, such as direct chemical attack and solvation.

Nonconductivity of electricity. This capacity reduces the problem of electrolytic corrosion of connected metallic piping.

Light weight; ease of use. These features simplify handling and installation with attendant cost savings and worker accident reduction. Their light weight also reduces dead-load forces on the structure.

Hydraulically smooth surfaces. Results in most efficient flow and minimal accumulation of deposits.

Elimination of brazing and soldering. During construction, fire hazards and work accident hazards from torches used for brazing and soldering of metal piping are eliminated.

Low stiffness. Allows for the accommodation of deformations with minimal stress development. For certain materials, i.e., PE and PB, stiffness is low enough to allow for coiling of smaller diameter pipe. Decreases sound transmission.

Ductility. The capacity of thermoplastics to undergo significant permanent deformation without fracture allows them to shed off, by means of localized deformation, concentrated earth and other loading.

Durability. Properly selected and installed thermoplastics piping can provide very high reliability and long service life.

Energy efficiency. Significantly less energy is required to manufacture a given length of plastic pipe than that needed to produce most metal pipes. For example, it has been estimated that the energy requirements for the manufacture of 2-inch, 160-pound-per-square-inch (psi) PE pipe are 60 percent of those for the same size copper pipe.

LIMITATIONS

The principal limitations of thermoplastics piping include:

Time, temperature, and environment sensitivity. All the engineering properties, including strength, stiffness, and strain at failure, are dependent on load, duration of loading, temperature, and environment. Furthermore, the sensitivity can vary not only from one type of plastic to another (i.e., PVC and PE), but also can significantly differ within the same material type depending on the exact choice of polymer and additives as well as on processing conditions. These limitations must be recognized in product design and standardization, particularly when limiting conditions, such as higher temperature, are likely to be encountered in actual service.

<u>High coefficient of expansion/contraction</u>. The thermal expansion/contraction of unrestrained plastics can be from 6 to 10 times greater than with metals. This requires greater attention to support spacing in horizontal runs. However, because of their significantly lower stiffness, expansion/contraction can often be restrained without the development of excess forces.

<u>Light weight</u>. Because of its low mass, thermoplastic piping offers little acoustical dampening to airborne sound. The light weight, in combination with low stiffness, can also result in greater movement of piping from hydraulic loads. This necessitates that more attention be given to proper pipe restraint in above-ground installations.

<u>Electrical insulator</u>. Grounding of electrical appliances and electrical circuits must be accomplished via a separate conductor and ground spike.

<u>Combustibility</u>. Control of the spread of fire, and toxic gases from plumbing chases, or walls containing thermoplastic materials, may require special attention to installation details, particularly when penetrating fire-rated building components. However, modern fire technology indicates that such attention to effects of wall penetration is appropriate to all piping materials.

DESCRIPTION OF THERMOPLASTIC PIPE MATERIALS

"Plastic" is as indefinite a term as "metal." First, plastics consist of two basic groups, thermosetting and thermoplastics, which are both used in the manufacture of piping. Thermoplastics, as the name implies, soften upon the application of heat and reharden upon cooling. This characteristic enables them to be readily formed into a wide variety of different shapes. Thermosetting plastics, on the other hand, form permanent shapes when cured by the application of heat or a "curing" chemical.

Secondly, among the thermoplastics the characteristics and properties depend both on the specific chemical composition of the base resin and on the kind and amounts of additives included in the commercial composition. For example, it is possible to formulate mixtures of polyvinyl chloride (PVC) resins plus appropriate additives to yield "vinyl" compositions that range from a clear, soft, and pliable product (such as is used to produce laboratory tubing and upholstery) to a rigid and strong material (such as for pressure pipe). For these reasons, standardization of plastic pipe materials not only classifies the base resin, but also defines the composition by a combination of specific end-property requirements that take into account possible property variability contributed by formulation additives of a plastics composition. Additives for plastics can consist of antioxidants, stabilizers, colorants, ultraviolet protective screens, lubricants, property modifiers (i.e., impact and stiffness), and fillers.

In the case of thermoplastic pipe materials intended for pressure piping and potable water service, the current material classification system requires that <u>each commercial composition</u> used for such product be evaluated to: (1) demonstrate that no deleterious substance will leach from the plastic into the potable water; and (2) establish the material's long-term strength under conditions of service. These requirements recognize that, because of variabilities in polymer and additives, plastics compositions cannot be adequately defined solely on the basis of the "short-term" physical property requirements that are part of ASTM material standards. These requirements alone are not a source of sufficient assurance that potable water quality will not be adversely affected and that the material has adequate long-term strength capacity for the intended service. Consequently, for a plastic material to be acceptable for potable water pressure piping, it must satisfy, under requirements in effect since the early 1960s, two additional criteria: (1) it must be approved for contact with potable water based on requirements not less strict than those in National Sanitation Foundation (NSF) Standard No. 14; and (2) it must have a recommended hydrostatic design stress (HDS) based on long-term pressure testing. The NSF requirements and programs are described in detail in a companion presentation by Dr. Nina McClelland of NSF.

The recommended HDS for individual commercial formulations is established from long-term pressure testing data on the basis of ASTM D 2837, "Standard Method for Obtaining Hydrostatic Design Basis for Thermoplastic Pipe Materials." The Hydrostatic Stress Board of the Plastics Pipe Institute (PPI) has been issuing since the mid-1960s recommendations of HDS for commercial grades of thermoplastic piping materials. These recommendations, which are based on the methods of D 2837 and the additional requirements given in PPI TR-3, "Policies and Procedures for Developing Recommended Hydrostatic Design Stresses for Thermoplastic Pipe Materials," appear in the regularly updated report PPI TR-4, "Recommended Hydrostatic Strengths and Design Stresses for Thermoplastic Pipe and Fittings Compounds." Nearly all U.S. thermoplastic pipe pressure standards reference the PPI recommendations. NSF policy requires such recommendation for each approved pressure piping material.

The distinctive characteristics of the principal thermoplastic piping materials are as follows:

<u>Polyvinyl Chloride (PVC)</u>. PVC piping is made from compositions containing no plasticizers and only minimal quantities of other ingredients. To differentiate these materials from flexible, or plasticized, PVCs (from which are made such items as upholstery, luggage, and laboratory tubing), they have been labeled rigid PVCs. Rigid PVCs used for piping range from a Type I to Type III as identified by a previous classification system that is still in use. In this system the Type designations are supplemented by Grade designations (i.e., Grade I or II) which further define the material's properties. Type I materials, from which most pressure and non-pressure pipe is made, have been formulated to provide optimum strength as well as chemical and temperature resistance. Type II materials are those formulated with modifiers that improve impact

strength but that also somewhat reduce, depending on modifer type and
quantity, the aforementioned properties of Type I materials. There is
little call for Type II pipe, as the impact strength of the stronger Type I
pipe is more than adequate for most uses. Type III materials contain some
inert fillers which tend to increase stiffness concomitant with some
lowering of both tensile and impact strength and chemical resistance. Some
non-pressure PVC piping, such as that used for conduit, sewerage, and
drainage, is made from Type III PVCs.

The currently used classification system for rigid PVC materials for
piping and other applications is described in ASTM D 1784, "Standard
Specification for Rigid Polyvinyl Chloride and Chlorinated Polyvinyl
Chloride Compounds." This specification categorizes PVC materials by
numbered cells that designate value ranges for the following properties:
impact resistance (toughness), tensile strength, modulus of elasticity
(rigidity), deflection temperature (temperature resistance), and chemical
resistance. The following table cross-references the designation of the
principal PVC materials from the older to the newer classifications system.

| CELL CLASSIFICATION SYSTEM OF | OLDER SYSTEM | |
ASTM D 1784 MINIMUM CELL CLASS	TYPE AND GRADE	DESIGNATION
12454-B	Type I, Grade I	PVC 11
12454-C	Type I, Grade II	PVC 12
14333-D	Type II, Grade I	PVC 21
13233	Type III, Grade I	PVC 31

The designation of PVC materials that have been formulated for
pressure piping also includes a code indicating their maximum recommended
HDS for water at 73° F as determined from long-term pressure testing. For
example: PVC 1120 is a Type I, Grade I PVC with a maximum HDS of 2,000 psi
for water at 73° F; and PVC 2110 is a Type II, Grade I PVC with a 1,000 psi
HDS. The last two digits in the number code identify, in hundreds of psi,
the material's maximum recommended design stress. Table 7 presents all the
PVC designations used in pressure piping.

The combination of good long-term strength with higher stiffness
explains why PVC has become the principal plastic pipe material for both
pressure and non-pressure applications. Major uses include: water mains;
water services; irrigation; drain, waste, and vent (DWV) pipe; sewerage and
drainage; well casing; electric conduit; and power and communications
ducts. A much broader range of fittings, valves, and appurtenances of all
types is available in PVC than in any other plastic.

Chlorinated Polyvinyl Chloride (CPVC). As implied by its name,
chlorinated polyvinyl chloride is a chemical modification of PVC. CPVC has
properties very similar to PVC, but the extra chlorine in its structure
extends its temperature limitation by about 50° F (28° C), to nearly 200° F
(93° C) for pressure uses and about 210° F (99° C) for non-pressure
applications. ASTM D 1784, the rigid PVC materials specification, also

covers CPVC which it classifies as Class 23477-B. By the older designation
system, it is known as Type IV, Grade I PVC. CPVC's for pressure pipe are
designated CPVC 4120 (i.e., Type IV, Grade I CPVC with a maximum
recommended hydrostatic design stress of 200 pounds per square inch
(lb/in^2) for water at 73.4° F. At 180° F (82° C) the maximum recommended
hydrostatic design stress for CPVC is 500 lb/in^2 (3.4 MPa).

Principal applications for CPVC are for hot or cold water piping and
for many industrial uses that take advantage of its higher temperature
capabilities and superior chemical resistance.

Polyethylene (PE). Polyethylene is the best-known member of the
polyolefin group (plastics that are formed by the polymerization of
straight-chain hydrocarbons, known as olefins), which includes ethylene,
propylene, and butylene. Polyethylene plastics are tough and flexible even
at subfreezing temperatures. They are generally formulated with only an
antioxidant (for protection during processing) and some pigment (usually
carbon black) or other agent designed to screen out ultraviolet radiation
in sunlight which, over long-term exposure, could be damaging to the
natural-color polymer.

ASTM D 1248, the PE molding and extrusion materials specification,
classifies these materials into three types depending on the density of the
natural resin. Type I consists of lower-density materials which are
relatively soft and flexible and have low heat resistance. Type II PEs are
of medium density, slightly harder, more rigid, and more resistant to
elevated temperatures; they also have better tensile strength. Type III
materials show maximum hardness, rigidity, tensile strength, and resistance
to the effects of increasing temperature. Pipe is made almost exclusively
from Type II and Type III PEs. ASTM D 1248 also provides for grade
designations to further classify PEs according to other physical
characteristics. PE piping materials for pressure piping are classified by
a designation system that combines the type and grade coding with that for
the maximum HDS for water at 73.4° F (23° C). PEs used for pressure piping
are listed in Table 6.

Outstanding toughness and relatively low flexural modulus, which
permits coiling of smaller diameter pipe, are large factors in PE's
prominence in gas distribution and water service piping. Other features,
such as heat fusibility, good abrasion resistance, and availability in
large diameters (up to 96 inches) account for the use of PE piping for
chemical transfer lines, slurry transport, sewerage force mains, intake and
outfall lines, and renewal (by insertion into the old pipe) of deteriorated
sewers, gas, water, and other pipes.

Polybutylene (PB). Polybutylene is a unique polyolefin because its
stiffness resembles that of low-density PE but its strength is higher than
that of high-density PE. However, its most significant feature is its
better retention of strength with increasing temperature. Its higher
temperature limit is higher than that of any PE: nearly 200° F for
pressure uses and somewhat higher for non-pressure applications.

PE piping materials are covered by ASTM D 2581. Pressure piping PBs
are designated as PB 2110 (see Table 7), which indicates a maximum design
stress of 1,000 psi for water at 73° F. At 180° F this design stress is
500 psi, the same value as for CPVC.

Major applications for PB piping take advantage of its improved
temperature resistance. They include hot/cold piping and industrial uses
such as hot/cold effluent lines. Because of its excellent abrasion
resistance, PB is also used for slurry lines.

Acrylonitrile-Butadiene-Styrene (ABS). ABS comprises a family of
materials that are formed from three different monomers (chemical building
blocks): acrylonitrile, butadiene, and styrene. The properties of the
components and the way in which they are combined can be varied to produce
a wide variety of properties. Acrylonitrile contributes rigidity,
strength, hardness, chemical resistance, and heat resistance; butadiene
contributes toughness; and styrene contributes gloss, rigidity, and ease of
processing.

ASTM D 1788 classifies ABS plastics into numbered cells that designate
value ranges for each of three properties: impact strength (toughness),
tensile stress at yield (strength), and deflection temperature under load.
ABS pipe materials are categorized into Types and Grades in accordance with
established minimum cell requirements for each Type and Grade. Like the
other major thermoplastics, ABS materials for pressure pipe are designated
by a coding that identifies both short-term properties and long-term
strength. ABS pressure-pipe materials are listed in Table 7.

An advantageous combination of toughness with good strength and
stiffness largely accounts for the most common uses of ABS pipe, i.e., for
DWV applications as well as for sewers, well casings, and communications
ducts.

A foam-core construction ABS pipe is commercially available. The wall
of the product consists of thin inner and outer solid skins sandwiching a
high-density foam. The primary benefit of this construction is improved
ring and longitudinal (beam) stiffness in relation to the amount of
material used. Applications of foam-core pipe are for non-pressure uses.
A standard for a PVC pipe with a foam-core wall construction is under
development at ASTM.

Polyacetal, also known as polyoxymethylene (POM), is a strong, hard,
highly crystalline thermoplastic offering good rigidity, strength, and
toughness, and relatively good temperature resistance. Because of these
properties, POMs are increasingly being used to mold various hot/cold water
plumbing components including valves, faucets, and compression fittings for
use with PB and CPVC tubing.

Crosslinked PE. A new standard has recently been issued by ASTM which
covers hot/cold water piping made from crosslinked PE (XPE). The
crosslinking changes the PE from a thermoplastic to a thermoset and endows

TABLE 7

<u>MAXIMUM RECOMMENDED HYDROSTATIC DESIGN STRESSES (RHDS)
FOR THERMOPLASTIC PIPE MATERIALS FOR WATER AT 73° F</u>

MATERIAL DESIGNATION[1]	MAXIMUM RHDS, IN PSI, FOR WATER AT 73° F[2]
PE 1404	400
PE 2305	500
PE 2306	630
PE 3036	630
PE 3406	630
PE 3408	800
XPE 330	630
PB 2110	1000
CPVC 4120	2000
PVC 1120	2000
PVC 1220	2000
PVC 2110	1000
PVC 2112	1250

[1]The first two digits code the material according to short-term properties. The last two digits code the maximum RHDS expressed in hundreds of pounds per square inch.

[2]Since thermoplastics, even though of the same material designation, may be affected differently by increasing temperature, RHDS for higher temperatures must be established for each specific commercial product.

the material with improved higher-temperature strength. Because they are
crosslinked, these materials cannot be heat fused. XPE has not yet been
commercially offered in the United States and it is not presently approved
by any major plumbing code.

Styrene Rubber (SR). Styrene rubber compositions are combinations of
polystyrene, in the greater part, with rubber. SR plastics are relatively
strong and stiff. However, their impact strength is somewhat lower than
other plastics, making SR more susceptible to damage by impact,
particularly during cold weather.

SR plastics are used exclusively for non-pressure applications. Once
relatively popular, they are now used little, having been largely displaced
by the other materials.

Solvent Cement Materials. Solvent cements, which are often used when
joining PVC, CPVC, and ABS piping, are compositions consisting essentially
of a solvent, or combination of solvents, in which has been dissolved a
quantity of the base plastic material sufficient to give the cement the
body and consistency required for proper applications. Small amounts of
inert fillers are sometimes also added to control these properties as well
as shrinkage during drying. The principal solvents used include
tetrahydrofuran (THF) for PVC and CPVC cements and methyl ethyl ketone
(MEK) for ABS cements. To control rate of drying and other properties,
PVC, CPVC, and ABS cements will sometimes include other solvents, most
often cyclohexanone (CH) and dimethylformamide (DMF). As in the case of
pipe, solvent cement compositions for use with potable water piping must be
approved for this purpose by NSF or an equivalent authority.

FUTURE TRENDS

The generally successful use of plastics piping over the past 30 years
has demonstrated that it is a cost-effective alternative for many water
distribution and plumbing applications. As with any new material, some
performance problems have occurred but in each case the underlying causes
have been identified and appropriate correction measures taken. None of
the noted problems suggests any inherent limitations by any of the plastics
piping systems to providing reliable and long-term service for the intended
applications. Enhancement in quality assurance requirements of current
standards and better attention to certain installation practices have
always been satisfactory resolutions to any noted performance inadequacies,
except, of course, for cases of clear misapplication of the product.

Whenever surveys have been conducted comparing the performance of
plastics to traditional piping materials, plastics have always shown up
well. For example, a recent survey conducted by the American Gas
Association indicates that during the year 1981 the reporting gas
distribution companies have experienced leaks in plastic piping at a rate
only 40 percent of that experienced for metallic piping, with no evidence

that any significant problems are indicated with currently available materials.

A survey during 1981 by the American Water Works Association (AWWA) Standards Committee on Thermoplastic Pressure Pipe revealed no reported problems with PVC C 900 pipe in water distribution. In an AWWA survey during 1982 for polyolefin water service pipe, 80 percent of the respondents replied that their experience with these materials shows them to be as good as, and more often than not better than, conventional materials. An investigation of the other 20 percent by a private consultant disclosed that improper installation and a problem with a limited production run of one pipe product accounted for a higher incidence of problems. Both of these causes have been addressed by the addition of appropriate quality assurance requirements and broader adoption of good handling and installation practices.

From the first introduction of plastics piping, producers and users have been alert to identify and correct problems with plastics. This is evident by the intense activity in such standard writing organizations as ASTM, AWWA, API, and NSF in enhancing plastic piping product standards by adopting the lessons of technology and experience to better define the end product and its proper application. The result is the steadily improving performance reliability and durability of plastics piping.

The primary obstacle to increased utilization of plastics is not any inherent lack of capability but rather the exploitation of public and worker concerns over safety and health risks that are often associated with synthetic materials. The following allegations have been offered in support of restricting the acceptance of plastics piping: The possible leaching of toxic substances that could be present in plastic materials into drinking water; health risks to the worker and user created by the use of solvent cements to join plastics piping; health risks by the possible permeation through the pipe wall of toxic substances that might be present in the soil in contact with buried piping; fire and accompanying toxic gas hazards posed by the combustibility of plastics; and decreased labor utilization because of the greater ease of installing plastics. Obviously, any concern dealing with public and worker health and safety deserves to be duly considered and any necessary precautionary measures must be incorporated into product standardization and utilization. For plastics such scrutiny and action, which shall be elaborated by others in this workshop, is a matter of record since the first use of these materials for piping. For example, the NSF requirements date back to the late 1950s.

All the objective scrutiny that plastics piping has been receiving since that time points to the conclusion that these materials, when properly used, present no significant health and safety risks to the public and to the installer. Consequently, there is no apparent basis for speculating that the acceptance of plastics in general, or any material specifically, may be curtailed by water quality/material problems. Rather, future trends in market share shall be largely influenced by technical developments in materials and products which will affect their ease of

application, performance, and ultimate best effectiveness. To this end
there is considerable activity by a number of plastics piping interests.

There is also considerable activity, as evidenced by this workshop, in
scrutinizing the effects of plastics and other materials on water quality.
Such balanced scrutiny is appropriate and welcomed. In the case of
plastics this might result in revision or additions to current material
requirements which, if needed, can readily be included into an existing
system, an opportunity that does not exist for the other piping materials.

Part II

Impact of Metallic Plumbing
Materials on Water Quality

IMPACT OF LEAD PIPING AND FITTINGS
ON DRINKING WATER QUALITY

Peter C. Karalekas, Jr.

Sanitary Engineer
Region I, Water Supply Branch
U.S. Environmental Protection Agency
Boston, Massachusetts 02203

HISTORICAL PERSPECTIVES

Lead pipe for conveying drinking water has been used for centuries (1). Among the known advantages of lead are its flexibility, durability, and long life. For these reasons, lead was used extensively in the United States beginning in the nineteenth century. Unfortunately, the hazards of using lead pipe, which were known at the time, did not deter many public water systems from using the material.

For example, in 1845 a report on water supplies for the city of Boston (2) concluded that:

Considering the deadly nature of lead poison, and the fact that so many natural waters dissolve this metal, it is certainly [in] the cause of safety to avoid, as far as possible, the use of lead pipe for carrying water which is to be used for drinking. The best substitute is found in a pipe well coated with pure tin on the interior -- such pipes being quite safe and in every way preferred.

However, this warning was not heeded, and in 1899 Clark (3) reported that in addition to Boston, 70 other communities in Massachusetts reported the use of lead or lead-lined service pipes. Commenting at length on Clark's work, Forbes (4), in an address to the members of the New England Water Works Association in 1900, concluded that Clark's work showed:

The many causes which make it possible for water to dissolve lead, and should teach us to look with extreme caution to the use of lead service pipes. It is also quite a serious matter, from a financial standpoint, to invest thousands of dollars in lead pipe, and then find that we have an element of serious danger in our midst which, sooner or later, must be remedied. The health of a community must first be considered; this goes without saying.

Now, in light of so much conclusive evidence which is within our
reach, what position should we take in this matter in constructing and
maintaining a system of water works? Is it not wiser and better to
eliminate all possible elements of danger from a thing so vital as the
water which we must daily take into our bodies? We must bear this
fact in mind; that ordinary citizens know little about these things,
and trust to those who have charge of the water supplies to furnish
them with a good and safe water and we should not be unmindful of the
confidence placed in us.

Again the warning was not heeded. In a survey of water utilities in
1924, Donaldson (5) reported that 51 percent of the 539 cities surveyed
across the country indicated they used lead or lead-lined services to some
extent.

More recently, two nationwide surveys have demonstrated that lead is
still a significant problem for some water supplies. In a study of 969
water supplies, McCabe et al. (6) concluded that 2 percent of the
population served was receiving lead in excess of the mandatory limit of
0.05 milligram per liter (mg/l). Patterson (7), in a survey of 580 cities,
found that 2.5 percent of the water samples exceeded 0.05 mg/l.

In a more recent survey (8) conducted by the U.S. EPA of 153 of the
largest water systems in 41 states, 112 systems indicated they had used
lead services during some period of time. However, at the time of this
survey only the City of Chicago continues to install lead services, as
mandated by their planning code. Lead goosenecks have also been widely
used and 94 systems indicated their use in the past.

MONITORING FOR LEAD IN DRINKING WATER

Current EPA regulations (9) for monitoring trace metals in drinking
water require that one sample per year be taken for lead at a consumer's
tap. In the author's opinion, a single sample per year is inadequate to
assess the concentrations of lead in drinking water from the corrosion of
lead pipe. Among the variables that have to be considered in a sampling
scheme to detect lead are contact time, water temperature, length of lead
pipe, and water quality. In order to assess the variation in lead
concentration in drinking water that had been standing for varying lengths
of time in piping, it was considered necessary to collect three samples at
each home. Samples were collected by the homeowners who followed the
instructions in Table 1.

Samples were collected at the kitchen sink the first thing in the
morning before any water was used in the home. Sample 1, the interior
plumbing sample, was collected immediately upon opening the faucet. This
represented water that had been standing overnight in the fixture and
interior plumbing serving the faucet. Sample 2, the service line sample,
was collected after the sample collector felt a temperature change in the

TABLE 1

SAMPLING INSTRUCTIONS

After 11:00 p.m., do not use the kitchen cold-water faucet until collecting the water samples the next morning. Using the following directions, in the morning, collect the water samples at the faucet before using any faucet or flushing any toilets in the house. Fill the provided containers to one inch below the top and put the caps on tightly.

Sample 1:   Open the cold water faucet, immediately fill bottle No. 1 and turn off the water. Recap this bottle.

Sample 2:   Turn the faucet on and place your hand under the running water, and immediately upon noticing that the water turns colder, fill bottle No. 2. Recap this bottle.

Sample 3:   Allow the water to run for three additional minutes and then fill bottle No. 3. Recap this bottle.

The man from the Environmental Protection Agency will stop at your house on the morning of ___________, to pick up the samples. if you do not expect to be home, please leave the samples outside the front door.

---

water from warm to cold. Since water would be expected to warm slightly
after standing in interior plumbing, this cold water would represent water
that had been standing overnight just outside the foundation of the house
in contact with the interior of the lead service line underground. Sample
3, the water main sample, was collected after the water was allowed to run
for several minutes. This represented water that would have a minimum of
contact time with the service line and interior plumbing.

The author feels that this method of sample collection is a realistic
representation of the range of lead concentrations which the consumer is
likely to find in the water he drinks. It is certainly conceivable that
the first sample of water might be drawn directly by the homeowner in the
morning and used to make juice or coffee, or drunk directly. The second
sample, representing the usually colder water standing in the service line,
could be drawn by a person trying to obtain the coldest water possible to
drink. The third sample, taken after running the water, represents water
that would be found more commonly during the day when water is used
frequently for flushing toilets and cooking.

WATER QUALITY RESULTS

In order to assess the contribution to lead in drinking water from
lead pipe, water samples were collected at consumers' taps supplied through
a lead gooseneck or service line in Boston, Chatham, and New Bedford,
Massachusetts, and Bennington, Vermont.

Although it is recognized there may be other contributors of lead,
such as lead/tin solder, brass or galvanized pipe, or bronze fittings in
the systems, all homes selected for sampling in Boston, New Bedford, and
Bennington had lead service lines and those in Chatham had lead goosenecks,
which are expected to be the major contributors of lead in these systems.

Table 2 lists raw water quality data for various parameters for each
of these systems. It should be noted that all are very low in alkalinity,
hardness, and pH, with lead concentrations in the sources of supply below
the detection limit of 0.005 mg/l.

Table 3 shows the quality of water at consumers' taps after that water
passed through the distribution system, the lead service line or gooseneck,
and the internal plumbing. In each system, significant numbers of samples
contained lead in excess of the Maximum Contaminant Level (MCL) of 0.05
mg/l. The lowest number of samples exceeding the standard occurred in
Chatham, where there were lead goosenecks of approximately 2 feet in
length. The other systems, where there were longer lead service lines
running from the water main in the street to the house, had a greater
proportion of samples in excess of the MCL.

In Bennington, where the water was extremely corrosive, 28 of the 30
samples exceeded the MCL. The percentage of samples containing lead

TABLE 2

RAW WATER QUALITY DATA*

	BOSTON	CHATHAM	NEW BEDFORD	BENNINGTON
pH	6.5	6.3	6.0	5.5
Hardness as $CaCO_3$	14	20	10	6
Alkalinity as $CaCO_3$	8	3	8	2
Sodium	5	11.9	6.5	1
Zinc	<0.02	<0.02	<0.02	<0.02
Copper	<0.02	<0.02	<0.02	<0.02
Lead	<0.005	<0.005	<0.005	<0.005

*All values except pH in mg/l

TABLE 3

DRINKING WATER QUALITY AT CONSUMERS'
TAPS BEFORE CORROSION CONTROL*

	BOSTON	CHATHAM	NEW BEDFORD	BENNINGTON
No. Homes Sampled	10	14	10	10
No. Samples	42	30	30	30
No. Samples Pb >0.05	28	4	14	28
No. Samples Pb >0.005	42	20	26	30
Avg. Pb Concentration	0.128	0.017	0.06	0.158
Highest Lead Concentration	0.870	0.098	0.26	0.475
pH	6.5	6.3	7.3	5.5

*All values except pH in mg/l

greater than the detection limit ranged from 66 percent in Chatham to 100
percent in Boston and Bennington, indicating that all water reaching the
tap was picking up some lead. The average lead concentration for all
samples in Boston, New Bedford, and Bennington was above the MCL.

The foregoing illustrates the effect of the corrosive nature of low pH
soft water on various lengths of lead pipe ranging from short, 2-foot lead
goosenecks to service lines as long as 200 feet.

REMEDIAL ACTION

As a result of the identification of the problem of corrosion of lead
pipe and the resulting high lead concentrations in drinking water, Boston,
New Bedford, and Bennington instituted corrosion control in an attempt to
reduce lead concentrations. Table 4 shows the results of corrosion control
in each of these three systems.

In Boston (10), the pH was raised to 8.2 using sodium hydroxide,
resulting in a significant decrease in average lead concentrations and in
the number of samples exceeding the lead MCL. There are still a large
number of samples exhibiting some pickup of lead. A similar situation
occurred in New Bedford, where the pH and carbonate alkalinity were raised
using sodium carbonate (soda ash). Again, significant reductions occurred
in average lead concentrations and in the number of samples exceeding the
MCL. In Bennington, corrosion control involved the addition of a
combination of sodium hydroxide and sodium bicarbonate to raise both the pH
and carbonate alkalinity. This appears to be the most effective treatment
in that the average lead concentration was reduced to the lowest level and
the number of samples with detectable levels was also the lowest. However,
like Boston and New Bedford, there are still a small number of samples
(three in each system) with lead greater than the MCL.

It would appear that the final solution to the problem involves both
corrosion control and systematic replacement of existing lead services over
time, giving priority to the worst-case situations, since corrosion control
alone, while highly effective in reducing lead, does not completely
eliminate the problem.

SUMMARY, CONCLUSION, AND RECOMMENDATIONS

1. Widespread use of lead service lines and goosenecks occurred in
    many water systems.

2. Many water systems have less than the optimum treatment for
    controlling lead corrosion.

3. High lead concentrations in all communities studied can be found
    if water samples are taken in a careful, systematic manner.

TABLE 4

DRINKING WATER QUALITY AT CONSUMERS'
TAPS AFTER CORROSION CONTROL*

ITEM	BOSTON	NEW BEDFORD	BENNINGTON
No. Homes Sampled	13	8	10
No. Samples	39	24	30
No. Samples Pb >0.05	3	3	3
No. Samples Pb >0.005	29	21	11
Average Pb Concentration	0.020	0.026	0.014
Highest Lead Concentration	0.134	0.118	0.096
pH	8.2	8.5	8.1

*All values except pH in mg/l

4.   The Interim Primary Drinking Water Regulations requiring only one
     sample per year are inadequate to identify the presence and
     amount of lead resulting from corrosion.

5.   The solution to the problem of lead involves systematic removal
     of lead materials, and water treatment to reduce corrosion.

6.   Although lead services are a major contributor of lead, other
     sources such as lead/tin solder must also be considered in any
     overall control strategy.

7.   The author recommends the use of lead as an acceptable material
     for conveying drinking water be removed from plumbing codes.

REFERENCES

1.   Frontinus, S.J. 1973. The Water Supply of the City of Rome. New
     England Water Works Association, Boston.

2.   Report of Commissioners, Appointed By Authority of the City Council to
     Examine the Sources From Which a Supply of Pure Water May be Obtained
     for the City of Boston. J.H. Eastburn, City Printer, Boston (1845).

3.   Clark, H.W. 1899. An investigation of the action of water upon lead,
     tin and zinc, with especial references to the use of lead pipes with
     Massachusetts water supplies. Annual Report of the Massachusetts
     State Board of Health, Boston, Massachusetts.

4.   Forbes, F.F. 1900. A very brief discussion of lead poisoning caused
     by water which has been drawn through lead service pipe. Journal
     NEWWA 15:58-60.

5.   Donaldson, N. 1924. The action of water on service pipes. Journal
     AWWA 11:649.

6.   McCabe, L.J., J.A. Symons, and G.G. Robeck. 1970. Survey of
     community water supply systems. Journal AWWA 62:670.

7.   Patterson, J.W. 1981. Corrosion in Water Distribution Systems.
     Office of Drinking Water, U.S. Environmental Protection Agency,
     Washington, D.C.

8.   Chin, D., and P.C. Karalekas, Jr. 1984. Lead production utilization
     survey of public water supply distribution systems throughout the
     United States. U.S. Environmental Protection Agency, Boston,
     Massachusetts.

9. *National Interim Primary Drinking Water Regulations*, EPA, *Federal Register* (December 24, 1975).

10. Karalekas, P.C., Jr., et al. 1983. Control of lead, copper, and iron pipe corrosion in Boston. Journal AWAA 75:93.

IMPACT OF COPPER, GALVANIZED PIPE, AND FITTINGS ON WATER QUALITY

Chester H. Neff

Principal Scientist
Illinois State Water Survey
605 E. Springfield
Box 5050, Station A
Champaign, Illinois 61820

INTRODUCTION

The stated purpose of this seminar is to review drinking water problems associated with various plumbing materials. This presentation will explore the impact of galvanized steel and copper piping materials on water quality. This is a concern of environmental scientists who are evaluating the health hazard associated with the corrosion of these materials in a drinking water system.

For a corrosion engineer, the first instinct is to investigate the reverse issue; what effect does the water quality have upon copper or galvanized pipe? Although this is a subtle difference, both viewpoints have a common focal point. Each is concerned with the chemical or physical reaction of a metal with its environment, a sample definition of corrosion. Corrosion engineering is the application of science and art to prevent or minimize corrosion damage economically and safely. Therefore, the corrosion engineer will select the best available piping material that is compatible with the water quality, while considering other factors such as cost, safety, and regulations. This includes the National Drinking Water Regulations, which establish Maximum Contaminant Levels (MCLs) for several metals found in copper or galvanized steel systems.

Copper and galvanized steel have found widespread use in potable water systems for many years. Experience has shown that both materials have good corrosion resistance to drinking water when properly selected and properly installed. Either can be subject to corrosion failures or contribute to the health hazards problem when abuses occur in designing the system or by installation procedures. The National Association of Corrosion Engineers (NACE) has published guidelines (4) for architects, design engineers, and plumbing contractors to call attention to the corrosion problems in these systems which they should be aware of.

Table 1 presents some of the constituents found in drinking water that define water quality. Many investigators have reported on the impact of the various constituents on the corrosion of copper piping materials (5-11), galvanized steel (12-21), and copper alloys (22-25). With the exception of lead, only in recent years has the impact of piping materials upon water quality been investigated to any extent. This presentation will review recent literature and report recent findings of the author and coworkers on the subject.

CORROSIVE QUALITIES OF DRINKING WATER

Some investigators have attempted to classify drinking waters according to their potential to be corrosive to piping materials. Campbell (5) classified waters by the water source and treatment rather than by chemical analyses. Types of water recognized by Campbell were: <u>deep well waters</u>, which are not usually corrosive to galvanized steel but may produce pitting in copper; <u>upland surface waters</u>, which are soft, have a low pH, and contain organic matter, and which are aggressive to galvanized steel and nonaggressive to copper; <u>slow sand filtered surface waters</u>, which contain a natural organic corrosion inhibitor for copper and are not corrosive to galvanized steel; and <u>lime-softened waters</u> with a high pH, which are nonaggressive to copper, but can be aggressive to galvanized steel in the absence of carbonate hardness. In similar matter, Obrecht and Myers (17) outlined fifteen categories for potable water, based on the corrosiveness and the scaling potential of the water supply. Calcium, silica, sulfate, dissolved oxygen concentrations, and temperature were employed to categorize the water. Lane and Neff (18) classified Midwest water supplies into five types, based on hardness, chloride and sulfate; pH; and alkalinity in a similar manner as Campbell.

Other investigators have attempted to employ predictive corrosion indices to identify waters that are aggressive. The Langelier Index (26) is the U.S. Environmental Protection Agency (EPA) recommended parameter for estimating corrosivity of public water supplies, although many corrosion engineers also employ the indices developed by Ryznar (27), Larson and Skold (28), Dye (29), McCauley and Abdullah (30), and others. These indices have been reviewed by Rossum and Merrill (31) and Singley (32).

The chemical constituents that are commonly cited by investigators as influencing the corrosivity of a water supply are calcium, alkalinity, pH, carbon dioxide, sulfate, oxygen, silica, and temperature. The unknown influences (organics, polyphosphates) and the interaction of all the constituents are not understood. None of the previously referenced papers provide information on the solubilization of metals from galvanized steel and copper pipe or their impact on water quality.

Lead concentrations exceeding the MCL value were reported in the Boston, Massachusetts, water supply by Karalekas et al. (33) and in Worcester, Massachusetts, by O'Brien (34) in the 1970s. Those studies,

Table 1.   Common Constituents Found in Drinking Water

<u>Dissolved minerals</u>

Calcium	Hypochlorite	Bicarbonate
Magnesium	Fluoride	Carbonate
Sodium	Phosphate	Sulfate
Potassium	Borate	Chloride
Barium	Hydroxide	Nitrate
Strontium	Iron	Silica
Trace Metals	Manganese	

<u>Dissolved gases</u>

Oxygen	Hydrogen Sulfide	Methane
Carbon Dioxide	Nitrogen	Ammonia

<u>Suspended Matter</u>

Corrosion Products	Mineral Deposits	Microbial Organisms

<u>Miscellaneous</u>

Unidentified Organics	Trihalomethanes

along with the development of analytical procedures capable of detecting
trace metal concentrations (e.g., atomic absorption) and the passage of the
National Safe Drinking Water Act, prompted several researchers to study the
impact of plumbing materials on water quality.

METAL CONCENTRATIONS FOUND IN SAMPLES FROM HOUSEHOLD PLUMBING

The metals associated with copper or galvanized steel piping materials
and fittings are zinc, copper, lead, tin, cadmium, and iron. Copper, zinc,
and lead concentrations found in samples taken from household taps are
presented in Table 2. All of the studies (33-39) referenced employed
procedures for determining running and standing sample metal
concentrations. Some of the studies were primarily concerned with lead
concentrations contributed by lead service lines, although the samples were
also exposed to the building plumbing materials.

Lead, copper, and zinc were found in concentrations that exceeded the
MCL value. As anticipated, this occurred most frequently in the standing
samples. Cadmium was not found in significant enough concentrations in
household plumbing to be considered a potential health problem. Tin
concentrations were not commonly reported and are not likely to be a
potential problem in drinking water. Iron concentrations frequently exceed
the MCL value in many water supplies as a natural occurring mineral and are
a factor in corroding galvanized steel plumbing.

Newly installed plumbing materials that have not formed protective
oxide films or mineral deposits are found to have very high zinc, copper,
and lead concentrations when in contact with water for long periods of
time. An example of this phenomenon occurred very inconveniently during a
cooperative corrosion study the Illinois State Water Survey (ISWS) is
conducting with EPA assistance. A corrosion test loop (Figure 1) was
installed with new copper and galvanized steel piping in a public water
supply to simulate normal household water usage. Approximately 40 feet of
galvanized steel pipe was installed by licensed plumbers ahead of
approximately 40 feet of copper tube employing 50 Sn:50 Pb solder
fittings. Chrome-plated brass sample valves, brass shut-off valves,
corrosion specimens, bronze water meter, and a PVC pipe nipple (to separate
materials) were also employed in the test loop. Water flow was controlled
by an electrically operated solenoid valve activated by a timer. The timer
was programmed to allow water to flow 5 to 10 minutes every hour for 16
hours and then to shut down the system for 8 hours. Standing and running
samples were taken after 7 to 8 hours of contact with the plumbing
materials. When this test loop was first placed in operation, the timer
failed and did not operate during the first 6 weeks. Two sets of samples
were collected during the 34-day period in which the test loop was flushed
one time after 15 days to collect running samples. Extremely high iron,
zinc, copper, and lead concentrations were found in both copper and
galvanized steel piping (Tables 3 and 4). Even after 5 minutes of
flushing, the running samples continued to have a high zinc concentration.

Table 2. Metal Concentrations Found in Samples from Household Taps

Location	No. of sites	Metal	Concentration (µg/L)			
			Standing		Running	
			Mean	Max.	Mean	Max.
Worcester, MA	9	Pb	270	1900	0	0
Bridgeport, CT	10	Pb	10	40	<5	--
Marborough, MA	9	Pb	14	250	10	--
Chatham, MA	10	Pb	17	98	15	--
New Bedford, MA	10	Pb	76	260	13	--
Seattle, WA	34	Cu	450	2050	120	1670
		Zn	1740	5460	150	1730
		Pb	39	170	5	22
		Fe	1400	5400	280	1200
Trondheim, Norway	18 cities	Zn	348	2125	126	--
		Cd	--	8.6	--	--
		Cu	164	1100	52	--
		Pb	9.7	110	2.7	--
Netherlands	10 cities	Zn	344	650	--	--
		Cd	<0.5	0.5	--	--
		Cu	647	1870	--	--
		Pb	81	180	--	--

SIMULATED CORROSION TEST LOOP
Water Meter
Shut-off Valve
M
Water Supply
1/2 in. Schedule 80, Galvanized Steel Pipe, 30 — 40 ft.
Threaded
Galvanized
Fittings
Sampling
Valve
2—1/2 in. Galvanized Steel Corrosion Testers
1/2 in. Schedule 80, PVC, 6 in. Length
1/2 in. Type L Copper Tube, 30 — 40 ft.
Sampling
Valve
2—1/2 in. Copper Corrosion Testers
Solder
Fittings
60 min.
Timer
24 hour
Timer
Solenoid
Valve
T
T
110 VAC
Waste

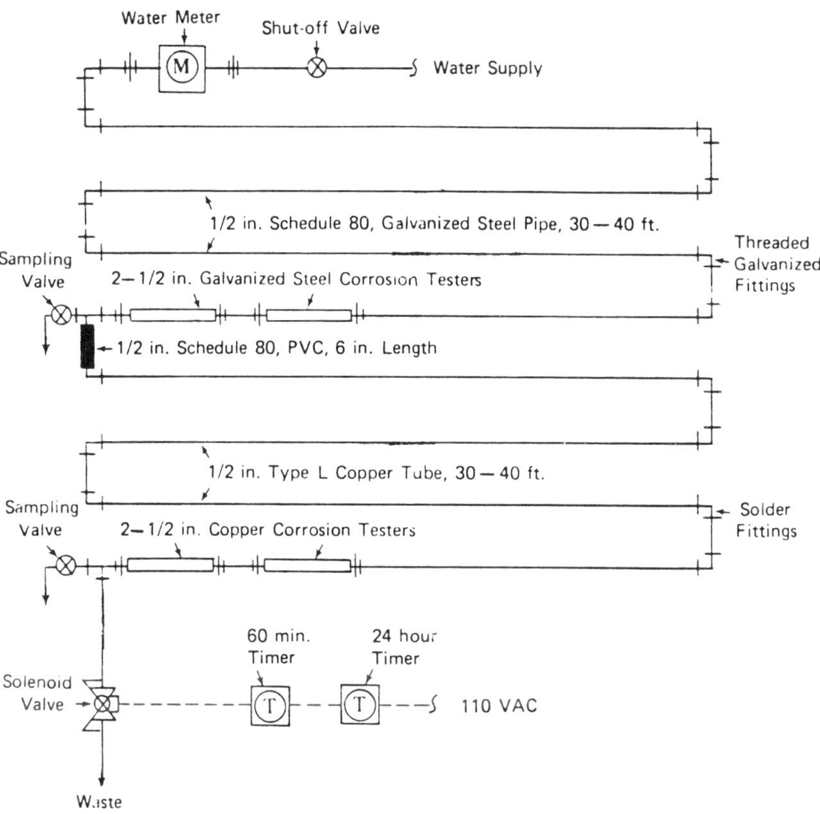

Figure 1

Table 3.   Total Metal Concentrations (mg/L) of Samples
from New Copper Plumbing*

Parameter		15 days, no flow	19 days, no flow	Normal operation
Standing	Pb	11.0	2.24	0.010
	Cu	1.687	0.343	0.005
	Zn	70.2	78.0	0.175
	Fe	14.3	7.43	0.13
	Cd	0.006	--	--
Flowing	Pb	1.89	0.330	0.004
	Cu	0.490	0.018	0.004
	Zn	17.4	31.0	0.090
	Fe	1.44	7.94	0.13
	Cd	0.001	--	--

* Note: Galvanized Plumbing Precedes Copper Plumbing

Table 4.   Total Metal Concentrations (mg/L) of Samples
from New Galvanized Plumbing

Parameter		15 days, no flow	19 days, no flow	Normal operation
Standing	Pb	0.770	0.710	0.001
	Cu	0.880	2.324	0.003
	Zn	67.60	11.2	0.150
	Fe	18.30	41.30	0.14
	Cd	0.009	--	--
Flowing	Pb	0.440	0.007	0.001
	Cu	0.013	0.005	0.003
	Zn	63.2	0.15	0.09
	Fe	2.66	0.11	0.12
	Cd	0.001	--	--

The metal concentrations increased again during the next 19 days of
exposure to the stagnant water in the loop. Normal operation (Tables 3 and
4) refers to the metal concentrations found in samples from the test loops
after the timer was replaced and programmed water flow was established.

This example dramatically indicates that extremely high lead, copper,
zinc, and iron concentrations can occur within newly installed household
plumbing. Such systems may require long flushing periods to reduce metal
concentrations below MCL values. Stagnant water in household plumbing will
frequently develop a bitter taste or turbidity due to the metal
concentrations. A consumer would probably flush such a system until the
water was palatable; however, the author is familiar with some galvanized
steel systems in which taste problems and high zinc concentrations
persisted for over a year after installation.

Plumbing materials do affect water quality. That impact is influenced
not only by the plumbing materials but by water quality, velocity,
temperature, exposed surface area, and system design. Most homes and
buildings are unique when these influences are considered, and the wide
variation of metal concentrations reported in the literature for drinking
water is not surprising.

IMPACT OF BRASS MATERIALS ON WATER QUALITY

The metal concentrations reporeted in Tables 3 and 4 exceed their
predicted solubility, which indicates suspended corrosion products were the
major component responsible for exceeding MCL values. Lead concentrations
were also seen as an indication that lead-tin solder was affecting water
quality. The chrome-plated brass sampling valves were also suspected as a
source of zinc, lead, and copper.

To determine the effect the brass sampling valves were having on
standing metal concentrations, a series of samples were taken from the
brass valve, which was then replaced with a high-density polyethylene
valve. Another series of standing samples were collected until the
corrosion test loop was permanently shut down. Table 5 shows the sudden
decrease in zinc, copper, and lead concentrations that occurred immediately
with the sampling valve change. Copper and lead concentrations were
reduced to near analytical detection limits. Zinc concentration decreased
to the background level contributed by the galvanized steel piping. The
chrome-plated brass sampling valve was determined by these tests to be a
major source of zinc, copper, and lead, even in the nonaggressive water
employed at this site. The sequential collection of 125- milliliter
samples also indicated that the brass sampling valve was influencing the
standing metal concentrations for more than 20 valve volumes of water.

The chrome-plated brass valves were employed in the ISWS-EPA corrosion
study to simulate the household sampling methods employed by the studies in
Table 2. Other brass valves with copper tubing and solder connections and

Table 5.   Influence of Valve Material on Total Metal Concentrations of Standing Samples

Date	Valve material	Zn (mg/L)			Cu (mg/L)			Pb (µg/L)		
		(a)	(b)	(c)	(a)	(b)	(c)	(a)	(b)	(c)
3/8	Chrome-plated brass	1.30	ND	ND	0.73	ND	ND	99	ND	ND
4/6	"	1.00	ND	ND	0.31	ND	ND	56	ND	ND
4/14	"	2.09	ND	ND	0.18	ND	ND	59	ND	ND
4/21	"	2.66	ND	ND	0.77	ND	ND	60	ND	ND
5/3	"	2.09	ND	ND	0.06	ND	ND	31	ND	ND
5/14	"	1.98	ND	ND	0.07	ND	ND	75	ND	ND
5/24	"	0.65	ND	ND	0.29	ND	ND	21	ND	ND
6/13	"	1.10	0.41	0.48	0.04	0.03	0.04	30	11	8
6/22	"	1.01	0.42	0.30	0.06	0.03	0.04	19	18	5
6/27	"	0.75	0.28	0.26	0.08	0.02	0.04	15	5	4
7/12	"	0.68	0.69	0.25	0.03	0.04	0.04	13	23	4
7/16	Polyethylene	0.30	0.20	0.24	<.01	<.01	<.01	1	1	1
7/21	"	0.22	0.16	0.14	<.01	<.01	<.01	1	1	1
8/5	"	0.35	0.26	0.22	<.01	<.01	<.01	2	1	2
8/15	"	0.55	0.22	0.01	<.01	<.01	<.01	1	1	1
8/22	"	0.46	0.21	0.17	<.01	<.01	<.01	1	1	1

a -- first 125 mL of water drawn from valve
b -- second 125 mL of water drawn from valve
c -- third 125 mL of water drawn from valve
ND -- not determined

more surface area may influence metal concentrations to a greater extent.
Samuels and Meranger (40) recently reported on metal concentrations leached
from eight types of new chrome-plated brass kitchen faucets by water of
varying quality. Data for three faucets are shown in Table 6. The authors
concluded that copper, zinc, chromium, cadmium, and lead were leached from
the faucets in varying amounts, depending on the type of faucet and the
water quality. Chromium and cadmium in the leachate were not considered
significant when compared to the MCLs for these metals. Werner (16)
observed that considerable quantities of copper and zinc were being
contributed by brass fittings and may be a significant factor in the
corrosion behavior of galvanized steel. These studies confirms ISWS
observations. Any attempt to correlate standing metal concentrations to
the corrosion of galvanized steel pipe, copper tube, lead service lines,
and solder should be preceded by isolating the system with plastic sampling
valves.

Brass or bronze water meters are also suspected to have similar metal
dissolution patterns as brass valves. Water meters are usually installed
in service line or piping near the entrace of the water into the
household. Standing or running metal concentrations may be influenced by
water meters for several system volume changes. The ISWS-EPA study will
attempt to determine if water meters are, in fact, contributing to the
metal concentrations of drinking water.

IMPACT OF MATERIALS VS. AGE OF SYSTEM

Metal concentrations found in samples exposed to stagnant water have
been discussed. This condition occurs only when homes are first built,
during vacation periods, etc. During normal period of water usage, metal
concentrations in samples are found to be much lower. Figure 2 illustrates
the change in zinc concentration with time of exposure. It also
demonstrates the difference in zinc concentration of established galvanized
steel plumbing (site 313) and new galvanized steel plumbing (site 315).
Zinc concentrations exceeded the MCL value for the first 9 months for the
new plumbing and required the entire 24 months of the study to attain
equivalent zinc concentrations found in the established system. In the
older system, zinc concentrations did not exceed the MCL for any sample
during the 24-month study and remained relatively constant for the entire
period. The decrease in metal concentration with time was also observed
for copper, lead, and zinc concentrations at other sites where new plumbing
materials were employed. The standing and running copper concentrations at
site 314, a copper test loop, are illustrated in Figure 3 showing the same
decline in copper concentration as seen for galvanized steel in this
particular water supply. The "spikes" or wide fluctuation in metal
concentrations was observed for all newly installed piping systems.

With the increased contact time of piping materials with water,
corrosion films may become more adherent and cover more surface area to
reduce metal concentrations found in samples. The "spikes" in

Table 6.   Total Metal Concentration Leached from Kitchen Faucets

Sample description			Faucet No. 1			Faucet No. 5			Faucet No. 6		
			Cu	Zn	Pb	Cu	Zn	Pb	Cu	Zn	Pb
Source	Leaching	pH	mg/L	mg/L	µg/L	mg/L	mg/L	µg/L	mg/L	mg/L	µg/L
Raw Water	1st	7.4	0.69	0.59	4.3	1.68	0.51	31.0	0.65	2.80	26.0
	2nd	7.4	0.83	0.40	3.4	0.43	0.65	20.0	0.41	1.70	24.0
Filtered Water	1st	6.3	1.82	2.75	1.4	0.78	1.15	55.0	2.47	2.55	26.0
	2nd	6.3	1.40	2.25	1.5	0.41	0.55	26.0	0.74	1.10	15.0
Treated Water	1st	8.4-8.6	1.03	0.40	0.7	0.37	0.39	49.0	3.33	4.70	45.0
	2nd	8.4-8.6	0.61	0.18	0.7	0.16	0.10	15.0	1.32	2.85	24.0
Well Water	1st	8.1	0.85	0.57	0.8	0.43	0.85	4.1	9.65	4.85	58.0
	2nd	8.1	0.28	0.46	0.4	0.30	0.51	1.8	3.54	2.00	45.0
Fulvic Acid (2 mg/L)	1st	6.2	0.19	0.55	4.3	0.65	0.37	110.0	1.76	1.32	18.0
	2nd	6.2	0.35	0.29	7.4	0.26	0.25	82.0	1.01	0.74	20.0

Re: Samuels and Meranger

SITE 313 AND SITE 315

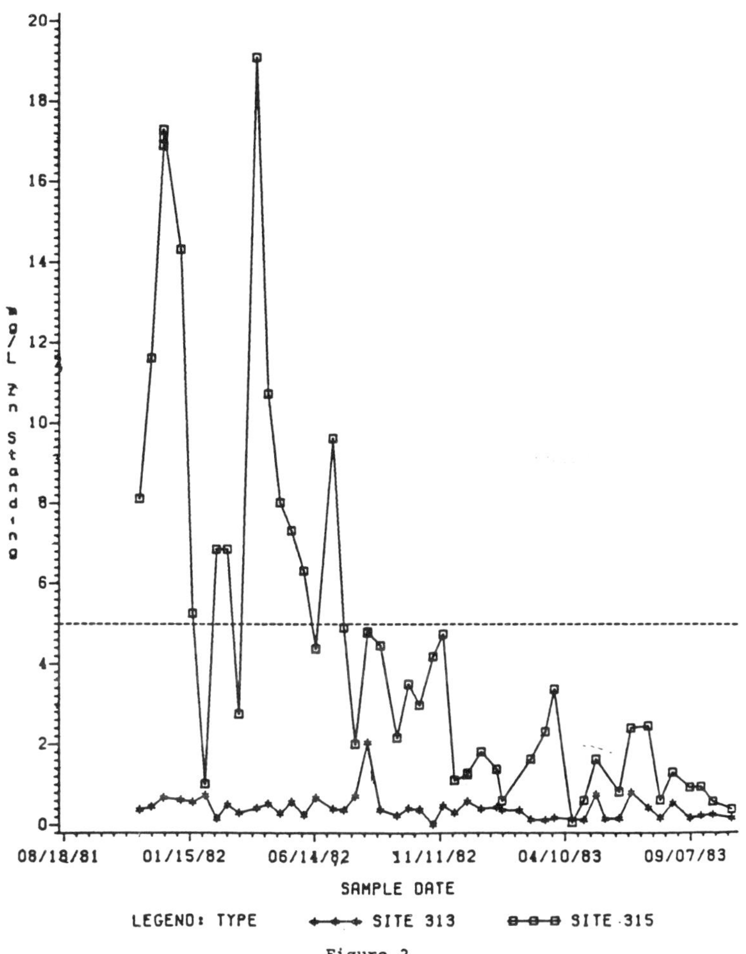

Figure 2

SITE 314

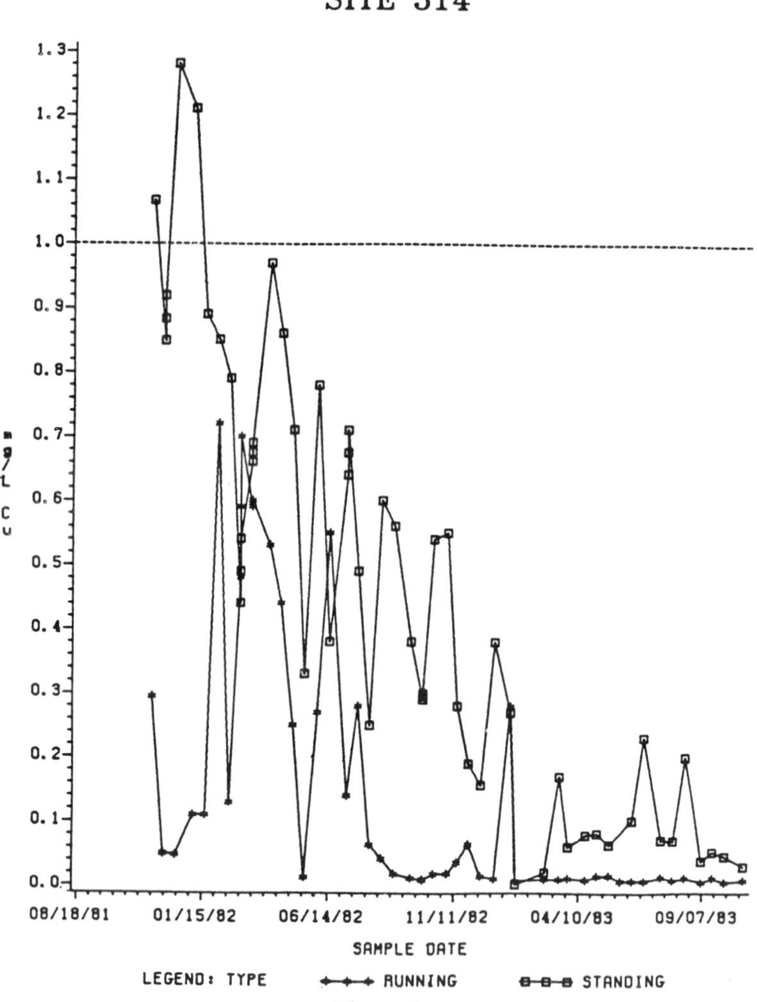

Figure 3

concentration values were assumed to be due to loose corrosion products
being transported through the plumbing. Contamination of drinking water by
zinc corrosion products in West Germany and referred to as "Zinkgeriesel"
has been reported by Werner (16). Serious corrosion damage of galvanized
steel has occurred in new buildings in which the corrosion products were
observed. Older buildings did not experience this type of damage.
Irregular zinc coating, faulty weld seam, faulty connections, residues from
construction, brass fittings, and temperature were noted as factors
responsible for "Zinkgeriesel."

NATURE AND SOURCE OF METAL CONTENT

Due to the "spikes" and some extremely high metal concentrations
observed at some test sites, four sites were selected to determine the
nature of metal contamination. A filtering procedure was implemented to
periodically determine soluble metal concentrations in addition to the
total metal concentrations of samples from each site. Tables 7 and 8
illustrate how the metal concentrations are influenced by sampling
sequence, filtration, and plumbing materials. The brass sampling valve
appears to be responsible for the zinc and lead concentrations in the
standing samples, although the galvanized steel test loop has a high
background zinc concentration. The galvanized steel pipe is the major
contributor of zinc which is more than 85 percent soluble at site 315.
Soluble zinc concentrations were observed to range from 1.0 to 2.6
milligrams per liter (mg/l) Zn.

Lead concentrations appear to be due to particulate lead, rather than
soluble lead, with the sources being the brass valves and lead-tin solder
of the copper plumbing. Standing copper concentrations are 17 micrograms
per liter (μg/l) in the galvanized plumbing with higher concentrations of
57 to 370 g/l observed in the copper plumbing. The copper tube is the
major source of copper at this site, while the contribution from brass
valves is not significant.

The major source of iron, manganese, and calcium concentrations
observed in Tables 7 and 8 is the water supply. Manganese and calcium are
present as a soluble species, whereas iron is present as both particulate
and soluble species. Sodium hexametaphosphate (5 mg/l) is being applied to
this water supply to control iron and manganese and may be influencing the
solubility equilibriums of metals at this site.

Another observation from these tables is that calcium, manganese, and
iron concentrations of standing samples are indicative of the water quality
at these sites 8 hours prior to the time the running samples were
collected. The practice of collecting standing samples prior to collecting
running samples infers that both samples have the same general chemical
characteristics, other than metal concentrations, which is often untrue.

Table 7. Metal Concentrations from Copper Plumbing (Site 314)*

	Sampling Procedure		Total metal concentrations (µg/L)					
Date	Flow condition	Filtered (0.4 µM)	Zn	Fe	Cu	Pb	Mn	Ca
2/24	Standing, 1st 125 mL	No	217	72	254	277	85	73
	2nd 125 mL	No	102	36	97	37	82	72
	3rd 125 mL	Yes	74	16	70	5	80	72
	Flowing, 10 min.	No	50	1480	13	1	224	118
	10 min.	Yes	50	830	12	1	211	118
4/28	Standing, 1st 125 mL	No	180	430	280	86	85	74
	2nd 125 mL	No	160	210	92	35	84	74
	3rd 125 mL	Yes	80	170	63	15	81	73
	Flowing, 10 min.	No	30	560	3	2	65	47
	10 min.	Yes	60	390	4	1	80	58
6/28	Standing, 1st 125 mL	No	280	2000	370	85	220	106
	2nd 125 mL	No	230	1680	150	15	230	107
	3rd 125 mL	Yes	150	380	57	7	180	106
	Flowing, 10 min.	No	60	890	7	32	30	35
	10 min.	Yes	30	270	5	7	40	34

* Note: Galvanized Plumbing Precedes Copper Plumbing

Table 8. Metal Concentrations for Site 315 (Galvanized Steel)

Date	Flow Conditions	Filtered (0.4 μM)	Zn	Fe	Cu	Pb	Mn	Ca
			Total Metal Concentrations (μg/L)					
2/24	standing, 1st 125 mL	no	1220	1360	11	90	99	73
	standing, 2nd 125 mL	no	1210	1060	17	61	100	72
	standing, 3rd 125 mL	yes	1040	240	5	7	93	71
	flowing, >10 min	no	60	1470	4	4	229	116
	flowing, >10 min	yes	60	590	4	2	203	115
4/28	standing, 1st 125 mL	no	1350	2010	7	194	98	76
	standing, 2nd 125 mL	no	1230	380	8	19	84	74
	standing, 3rd 125 mL	yes	1170	230	3	8	80	76
	flowing, >10 min	no	50	540	3	2	70	58
	flowing, >10 min	yes	30	370	2	2	80	63
6/28	standing, 1st 125 mL	no	2960	2250	8	18	240	107
	standing, 2nd 125 mL	no	2880	1480	8	8	240	106
	standing, 3rd 125 mL	yes	2640	180	6	2	210	107
	flowing, >10 min	no	90	1440	7	3	60	38
	flowing, >10 min	yes	50	540	6	2	80	45

IMPACT OF DESIGN AND INSTALLATION PROCEDURES ON WATER QUALITY

The selection of the proper piping materials is the most important single factor in designing a potable water system. The experiences of Campbell (6), Obrecht and Myers (17), Cohen (10), Lane and Neff (18), and others provides some guidelines for this selection; however, local experience should weigh heavily on the selection process. Although the proper material may be selected, poor pipe quality can be responsible for rapid failure. Galvanized steel pipe in recent years has had frequent occurrences of defective or inadequate zinc coatings. The quality of copper tube has been consistent, although carbon films formed in the manufacturing process have been reported to cause pitting. Piping should be inspected prior to installation. Perhaps tighter piping specifications and quality control procedures are required for materials exposed to drinking water.

Poor plumbing practices or "shoddy" workmanship can also result in corrosion of plumbing materials. It is not uncommon to observe dissimilar metals joined together, excessive threading of pipe ends, excessive use of flux and solder, inadequate reaming of pipe ends, or other manifestations of poor plumbing practice. Each of these factors can contribute to copper, zinc, or lead content of potable water, although the quantities contributed are not known.

A contributing factor to the corrosion of materials is velocity, which can be controlled by design. Copper tube is more susceptible to the erosion-corrosion attack by water than is galvanized steel. This type of corrosion is seen more often in recirculating water systems which have oversized pumps but has occurred in undersized copper tube in potable water systems. Unreamed tube ends may also cause flow disturbances that erode otherwise protective films in the vicinity of the fittings. It is not clear whether less adherent corrosion products or soluble copper is being generated by this corrosion process. Water velocity can influence corrosion rates and increase the transport of zinc corrosion products in galvanized steel plumbing.

Grounding of electrical circuits to water pipes in houses has been reported (41) to cause copper corrosion with resulting blue-water and copper concentrations of 6.5 mg/l. Corrosion by stray DC currents is well documented; however, stray AC current corrosion has rarely been documented. AC current corrosion has been reported (42) to increase with decreasing frequency; and at 5-Hz AC currents, corrosion is 1 percent of the equivalent DC current. Reliable information on the subject is not available; further investigation is warranted.

IMPACT OF WATER TREATMENT PRACTICES

A brief discussion is warranted on the water supply operations and maintenance program that can increase the corrosivity of the water toward

copper or galvanized steel. Chemical treatment programs for the purpose of
corrosion control, we will assume, are reducing the corrosion.

CONTROL OF WATER SUPPLY QUALITY

The water quality of public water supplies has been found to vary
drastically over short time spans during the ISWS-EPA corrosion study.
These fluctuations in water chemistry have been due to equipment failures,
operation procedures, disinfection procedures, treatment change, and
seasonal variations. Public water supplies are dynamic systems where
chemical equilibriums control the solubility of metals and corrosion
mechanisms. Any change in temperature, pressure, or ionic concentration,
in an otherwise stable system, will influence corrosion. One example is a
small community in Central Illinois which employs five wells for a source
of water. Well operation follows a schedule which pumps one well for one
day, another well the next day, and three wells the following day. The
quality of the water from these wells is significantly different, as noted
by sulfate, chloride, and hardness concentration ranges of 77 to 590 mg/l,
51 to 404 mg/l, and 265 to 364 mg/l, respectively. Without elaborating
further, water plant operators in small communities need information and
assistance to improve their awareness of operating practices which may
upset a stable system (if one exists).

CHLORINATION IMPACT

Another significant change observed for water supplies in Illinois
during the last 10 years has been a decision by many water suppliers to
abandon lime-softening and by other water suppliers to change to free
residual chlorine disinfection methods. Economic reasons have probably
been responsible for the change from lime-softening to clarification-only
treatment, while the Illinois EPA has encouraged free residual chlorination
methods. We have observed 2 to 4 mg/l free residual chlorine in some
supplies in order to maintain minimum residuals in the extremities of the
system. Application of sodium silicate and increased pH reduced the
corrosion that had been caused by high, free chlorine residuals in a copper
system of a state mental health facility. At another facility, pitting
failures of copper heat exchangers were reduced by lowering free chlorine
residuals from 2 mg/l to 1 mg/l. Increased solubilization of copper by
chlorine, which is pH dependent, has been cited in the literature (43).

POLYPHOSPHATE TREATMENT

The usage of polyphosphates to sequester iron has found widespread
acceptance among small water suppliers, since this alternative is more
economical than other iron removal processes and has the added benefit of
removing tuberculation from cast iron piping. It is this latter ability to
solubilize corrosion products which concerns the author.

Early reports (44,45) on the effect of polyphosphates on lead pipe
concluded that polyphosphates inhibited lead solubility below pH = 7.0 but

Table 9. Mean Metal Concentrations (μg/L) Observed in Copper Plumbing
(approx. 45 samples per site)

Site	Water Supply	Plumbing Age	CuR	CuS	FeR	FeS	PbR	PbS	ZnR	ZnS
301	A	new	37	267	89	215	14	106	166	403
303	A	old	7	25	53	57	1	3	26	85
305	B	new	25	188	60	142	1	17	24	84
307	B	old	8	37	50	53	1	2	20	22
308	C	new	489	1327	144	287	4	42	1116	2358
310	C	new	555	1277	164	134	5	10	1678	4331
312	D	new	258	675	177	759	30	170	24	202
314	E	new	125	419	1107	1408	455	2125	207	348
316	F	new	15	287	51	93	14	183	25	379
318	F	new	10	95	56	58	4	70	66	597

R -- running samples   S -- standing samples

Table 10. Mean Metal Concentration (μg/L)
Observed Galvanized Steel Plumbing
(approx. 45 samples per site)

Site	Water Supply	Plumbing Age	CuR	CuS	FeR	FeS	PbR	PbS	ZnR	ZnS
302	A	old	8	11	67	109	1	6	75	134
304	A	old	3	38	54	100	1	3	57	122
306	B	new	9	89	78	145	1	12	90	411
309	C	new	538	766	701	882	13	17	2214	5918
311	C	new	737	1411	244	677	6	15	3341	9845
313	D	old	27	79	150	150	4	8	156	418
315	E	new	10	39	1025	3009	5	213	365	4359
317	F	new	4	51	135	150	3	16	251	4376
319	F	new	5	16	70	320	3	80	286	6685

R -- running samples   S -- standing samples

slightly increased lead solubility above pH = 8.8. One author (45)
concluded: "Since the amount of lead taken up by waters of high pH value
is relatively small, the increase due to the addition of metaphosphate does
not seem to indicate any serious danger to public health, except possibly
in softened waters which employ metaphosphate for recarbonation."

In this paper, the reported standing lead concentrations from lead
pipe exposed at pH = 8.8 increased from 0.33 mg/l to 0.79 mg/l Pb as the
metaphosphate dosage was increased from 0 mg/l to 10 mg/l. To this author,
those lead concentrations appear significant, as are some lead values
observed in the ISWS-EPA study.

Preliminary mean metal concentrations are reported in Table 9 (copper
systems) and Table 10 (galvanized steel systems). Water supplies D and E
contained 1.0 and 4.0 mg/l polyphosphate (as PO_4), while water supply A
temporarily contained 0.5 mg/l polyphosphate. These three water supplies
have been exceeding the lead MCL value seriously at site 314, causing
concern that the polyphosphate may be increasing the solubility of lead
from lead-tin solder or brass fittings. Continued research on the effects
of polyphosphates on plumbing materials is required due to the potential
health hazard of these effects.

SUMMARY

Many studies have shown that copper tubing, galvanized steel pipe, and
brass fittings have increased the lead, zinc, iron, and copper
concentrations in drinking water. Water quality and water treatment will
influence the degree of impact these plumbing materials will have. The
MCLs for these metals have been exceeded most often in standing samples
from household taps. Brass sampling valves may have made a significant
contribution to the metal concentraions reported in many studies. Newly
installed copper or galvanized steel plumbing may require several months to
attain stable, minimum metal values in the drinking water. Stability may
never be achieved in some water supplies due to fluctuating water
chemistry. Disinfection methods and polyphosphate usage may also increase
the solubility of metals from copper or galvanized plumbing.

Continued research is required to identify the sources of these metals
and to control their solubility in drinking water. Basic information is
needed on the nature of protective corrosion films, the influence of
complexing agents on the film, and the water chemistry necessary to form
the films.

REFERENCES

1. EPA. 1976. National Interim Primary Drinking Water Regulations.
    EPA-570/9-76-003. U.S. Environmental Protection Agency, Office of
    Water Supply, Washington, D.C.

2. EPA. 1980. National Interim Primary Drinking Water Regulations, Amendments. _Federal Register_ 45(168):57332.

3. EPA. 1979. National Secondary Drinking Water Regulations. _Federal Register_ 44(140):42195.

4. NACE. 1983. Prevention and Control of Water-Caused Problems in Building Potable Water Systems (TPC-7). NACE Publications 52186. National Association of Corrosion Engineers.

5. Campbell, H.S. 1971. Corrosion, water composition, and water treatment. Proceedings, Soc. Water Treat. & Exam. 20:11.

6. Campbell, H.S. 1954. The influence of the composition of supply waters, and especially of traces of natural inhibitor, or pitting corrosion of copper water pipes. Proceedings, Soc. Water Treat. and Exam. 8:100.

7. Lucey, V.F. 1967. Mechanism of pitting corrosion of copper in supply waters. Brit. Corrosion Jour. 2:175.

8. Cruse, H., and R.D. Pomeroy. 1974. Corrosion of copper pipes. Jour. AWWA 8:479.

9. Obrecht, M.F., and M. Pourbaix. 1969. Corrosion of metals in potable water systems. Proceedings, 3rd International Congress on Metallic Corrosion, Vol. IV, Moscow.

10. Cohen, A. 1978. Copper in potable water systems. Heating, Piping, and Air-Conditioning 5:81.

11. Kenworthy, L. 1943. The problem of copper and galvanized iron in the same water system. J. Inst. Metals 69:67.

12. Britton, S.C. 1936. The resistance of galvanized iron to corrosion by domestic water supplies. J. Soc. Chem. Industry 1:19.

13. Kenworthy, L., and M.D. Smith. 1944. Corrosion of galvanized coatings and zinc by waters containing free carbon dioxide. J. Inst. Metals 70:463.

14. Cox, G.L. 1931. Effect of temperature on the corrosion of zinc. Ind. Engr. Chem. 23:902.

15. Slunder, C.J., and W.K. Boyd. 1971. Zinc: its corrosion resistance. Zinc Institute Publication.

16. Werner, G. 1984. Galvanized Pipes in House Installations, Zweckverband Landeswasserversorgung, West Germany.

17. Obrecht, M.F., and J.R. Myers. 1973. Potable water systems in buildings: deposit and corrosion problems. Heating, Piping, and Air-Conditioning, May.

18. Lane, R.W., and C.H. Neff. 1969. Materials selection for piping in chemically treated water systems. Materials Protection 8(2):27.

19. Gilbert, P.T. 1948. The corrosion of zinc and zinc-coated steel in hot waters. Sheet Metal Industries, October.

20. Werner, G., E. Wurster, and H. Southeimer. 1973. Corrosion tests of galvanized steel pipe by the Ground Water Supply Administration. GWF-Wasser/Abwasser 114.

21. Bachle, A., E. Dessiner, H. Weiss, and I. Wagner. 1981. The corrosion of galvanized and unalloyed steel pipes in drinking water of different hardness and neutral salt content. Werkstoffe und Korrosion 32.

22. Geld, I., and McCaul, C. 1975. Corrosion in potable water. Jour. AWWA 10:549.

23. Turner, M.E.D. 1961. The influence of water composition on the dezincification of duplex brass fittings. Proc. Soc. Wtr. Treat. Exam. 10.

24. Wormwell, F., and T.J. Nurse. 1952. The corrosion of mild steel and brass in chlorinated water. J. Appl. Chem. 2, Dec.

25. Larson, T.E., R.M. King and L. Henley. 1956. Corrosion of brass by Chloramine. Jour. AWWA 48, Jan.

26. Langelier, W.F. 1936. The analytical control of anti-corrosion water treatment. Jour. AWWA 28:1500.

27. Ryznar, J.W. 1944. A new index for determining amount of calcium carbonate scale formed by water. Jour. AWWA 36:472.

28. Larson, T.E., and R.V. Skold. 1958. Laboratory studies relating mineral quality of water to corrosion of steel and cast iron. Corrosion 14(6):285.

29. Dye, J.F. 1952. Calculations of the effect of temperature on pH, free carbon dioxide, and the three forms of alkalinity. Jour. AWWA 44(4):356.

30. McCauley, R.F., and M.O. Abdullah. 1958. Carbonate deposit for pipe protection. Jour. AWWA 50:1419.

31. Rossum, J.R., and D.T. Merrill. 1983. An evaluation of the calcium carbonate saturation indexes. Jour. AWWA, Feb.

32. Singley, J.E. 1981. The Search for a Corrosion Index. Jour. AWWA 73(11):179.

33. Karalekas, P.C., C.R. Ryan, C.D. Larson, and F.B. Taylor. 1978. Alternative methods for controlling the corrosion of lead pipe. New Eng. Water Works Assoc. 92(2):159.

34. O'Brien, J.E. 1976. Lead in Boston water: its cause and prevention. New Engl. Water Works Assoc. 90(2):172. ·

35. Zoeteman, B.C.J., and B.J.A. Haring. 1978. Introduction of Chemical Compounds into Drinking Water During Distribution. Ryksinstituut Voor Drinkwatervoorziening Mededeling 78-6.

36. Stegavils, K. 1975. An investigation of heavy metal contamination of drinking water in the City of Trondheim, Norway. Bull. Environm. Contam. and Toxicol. 14(1):57.

37. Haring, B.J.A., and B.C.J. Zoeteman. 1980. Corrosiveness of drinking water and cardiovascular disease mortality. Bull. Environm. Contam. Toxicol. 25:658.

38. Sharrett, A.R., A.P. Carter, R.M. Orheim, and M. Feinlieb. 1982. Daily intake of lead, cadmium, copper, and zinc from drinking water: The Seattle study of trace metal exposure. Environmental Research 28:456.

39. Sharrett, A.R., R.M. Orheim, A.P. Carter, J.E. Hyde and M. Feinleib. 1982. Components of variation in lead, cadmium, copper, and zinc concentration in home drinking water: The Seattle study of trace metal exposure. Environmental Research 28:476.

40. Samuels, E.R., and J.C. Meranger. 1984. Preliminary studies on the leaching of some trace metals from kitchen faucets. Water Res. 18(1):75.

41. Guerrera, A.A. 1980. Grounding of electric circuits to water sources: one utility's experience. Jour. AWWA 2:82.

42. Shreir, L.L. 1963. Corrosion. Vol. 2, Corrosion Control. John Wiley and Sons.

43. Atlas, D., J. Coombs, and O.T. Zajicek. 1982. The corrosion of copper by chlorinated drinking waters. Water Res. 16:693.

44. Hatch, G.B. 1942. Inhibition of lead corrosion with sodium hexametaphosphate. Jour. AWWA 33:85.

45. Moore, E.W., and E.E. Smith. 1943. Effect of sodium hexametaphosphate on the solution of lead. Jour. AWWA 34(9):1415.

SUMMARY OF IMPACT OF METALLIC SOLDERS
ON WATER QUALITY

Norman E. Murrell

H2M Consulting Engineers
125 Baylis Road, Suite 140
Melville, New York 11747

BACKGROUND

An incident of lead poisoning of a dentist's son in Smithtown, New York, prompted the consumer's request for testing of his home water supply. Water as a source of lead was discounted at first because the Water District had a 15-year record of lead-free distribution samples. Nevertheless, first-draw (zero to one minute) sampling showed lead in excess of the Drinking Water Standard of 50 micrograms per liter (μg/1). After three minutes of flow, sampling showed water well under the standard limit.

Lead solder was suspected as the source of the lead. Every home in the 8-year-old subdivision was tested. The two occupied homes with high first-draw lead levels both had recent plumbing additions using lead solder. A new home under construction had an even higher first-draw lead value of 7,100 μg/1.

Consulting engineers H2M/Holzmacher, McLendon & Murrell, P.C., conducted this testing and, at the same time, initiated an American Water Works Association (AWWA) literature search for studies of lead solder. Prior studies in the United States, Canada, and Europe showed high lead values in first-draw water after periods of non-use. The lead values were higher in water systems with low pH and soft water. Further studies were then conducted by H2M in Nassau and Suffolk counties, New York, to investigate the occurrence of lead leaching from new and old lead solder joints.

PRELIMINARY INVESTIGATIONS

In the Smithtown Water District, at the site of the original complaint, a time series test was run after a 4-hour period of nonuse with a first-draw value of 300 μg/1. It took about 48 seconds for the lead

value to drop below the drinking water standard. In the home under
construction, the first-draw lead value was 25,000 g/l and, after 80
seconds of flushing, the lead value still exceeded the drinking water
standard. In a Melville office building with 23-month-old plumbing, first
draw after a 71.5-hour shutdown (long weekend) was 120 g/l. The highest
level (200 g/l at 1.5 minutes) was calculated to coincide with the
location of lead-soldered joints within the plumbing system.

Continuing the investigation, more than 200 first-draw tests were
conducted by H2M in Nassau and Suffolk counties -- a 1,200-square-mile area
dependent on naturally low pH and generally soft groundwater for its water
supply. There were many results above the drinking water standard. County
health departments, the 13 towns in the bi-county area, and water suppliers
were informed of these findings. Independent testing by the Suffolk County
Water Authority, Suffolk County Health Department, Nassau County Health
Department, and Town of Hempstead Water Department have confirmed H2M's
findings (see Table 1).

In January 1983, a meeting was held with U.S. Environmental Protection
Agency (EPA) personnel in Cincinnati, Ohio, to discuss the investigations
held to date and to suggest further research focusing on the variables of
age of plumbing and the pH and hardness of the water supply. The South
Huntington Water District agreed to participate and a proposal for a
cooperative study was submitted to EPA in March 1983. The "Lead Solder
Aging Study" was approved by EPA to commence October 1, 1983.

EPA/SOUTH HUNTINGTON WATER DISTRICT COOPERATIVE STUDY

As of May 1984, 90 sites have been identified and selected for this
study, the South Huntington Water District portion of the low pH sampling
has been completed, and much raw data has been gathered but not yet fully
analyzed. Ten homes were selected in each of nine age groups from zero to
20+ years old. These sites would be sampled three times at different pH
levels: less than 6.2, 7.2, and 8.2. The homes were randomly selected to
obtain a geographic distribution in the 19.4-square-mile service area of
the District. The type of existing solder was verified in each home
through scrapings of an exposed solder joint and testing by atomic
absorption spectrophotometer. Of 95 homes so tested, only one had less than
0.50 percent lead in the solder; 67.3 percent of the homes had lead content
ranging from 55 to 65 percent in the solder.

First-draw samples were tested for copper and cadmium in addition to
lead. An additional sample was drawn for determination of various water
quality parameters including pH. Copper values above the 1 milligram per
liter (mg/l) Secondary Maximum Contaminant level (up to 7.77 mg/l at 6.9 pH
and 0 amps) were found. These sites were checked for stray electric
currents in the water service pipe to determine whether or not the high
copper values are caused through grounding of electric systems to the water
plumbing system.

TABLE 1

NASSAU COUNTY HEALTH DEPARTMENT TESTS

LOCATION	LEAD (μg/l)	pH	HARDNESS (mg/l)	% LEAD SOLDER	AGE (YEARS)
Locust Valley	17,000	7.0	41	60.3	0
Port Washington	4,400	6.8	49	61.7	1
Manorhaven	3,500	6.8	49	47.9	0
Woodbury	2,900	7.3	64	58.4	0
North Port Washington	930	7.0	49	50.3	0
North Hills	750	6.7	23	56.2	0
North Hills	530	6.7	23	60.0	0

TOWN OF HEMPSTEAD WATER DEPARTMENT TESTS

WATER DISTRICT	BUILDING AGE	TIME BEFORE FIRST DRAW	LEAD (μg/l)	
			0 MIN	1 MIN
East Meadow	8 months	8 hours	2010	7
East Meadow	8 months	24 hours	1500	240
Uniondale	3 years	8-10 hours	1049	5
East Meadow	8 months	8-10 hours	810	5
East Meadow	8 months	24 hours	760	380
Roosevelt Field	6 months	12 hours	569	385
Roosevelt Field	6 months	12 hours	338	5
East Meadow	3 months	24-48 hours	186	174

During low pH sampling of 64 first-draw tests for cadmium, only one sample at 42.5 μg/l was above the drinking water standard. Since the second sample at 10 seconds contained only 3 μg/l cadmium, we assume the presence of cadmium was caused by the faucet.

FIRST-DRAW LEAD FINDINGS

Of the first-draw lead samples taken, 61.84 percent have been above the drinking water standard. The results ranged from a low of 2.0 μg/l in a 1968 home at pH = 5.9, to values ranging from 500 to 1,200 μg/l (pH ranges from 5.6 to 6.9) at 10 homes constructed between 1981 and 1983. However, the highest lead value (1,300 μg/l) was obtained at pH = 5.9 in a 1968 home.

With only the initial findings at a low pH, there appears to be a correlation between the age of the lead solder and its ability to leach lead into the drinking water. Ignoring the first zero-second flush and allowing the first 300 milliliters (ml) of flush for the faucet's leaching of lead, the second sample at 10 seconds was compared for 75 homes. For homes constructed in 1980, 1981, 1982, and 1983, the second 125-ml sample (taken in March and April, 1984) indicated that 23 homes, or 70 percent, exceeded the 50- g/l drinking water standard, and 10 homes were less than this maximum contaminant level. For homes constructed between 1955 and 1979, only seven, or 17 percent, exceeded the drinking water standard at a low pH with 35, or 83 percent, less than this maximum contaminant level.

pH AND HARDNESS

Additional investigations have been undertaken to determine locations outside Long Island with water conditions that may be conducive to lead leaching. Research throughout the world indicates that leaching of lead is most acute in areas with low pH and soft water. This condition occurs not only on Long Island, but in many locations in New York State and throughout the nation. In New York State, 2,802 samples from 784 community water systems have been gathered as part of the Water Resources Investigation of the Chemical Quality of Water from Community Systems conducted by the U.S. Geological Survey and the New York State Department of Health. Review of this data shows many locations in New York City and throughout the state with pH and hardness values which may be conducive to lead leaching.

HEALTH EFFECTS OF LEAD

Adverse effects of lead on infants have been well documented, including anemia, colic, encephalopathy, nephropathy, and neuropathy. Reduced I.Q. has been associated with high lead levels. It has been recommended in at least two research studies of lead in drinking water that the safe value of lead for infants should be reduced from 50 μg/l to 25 μg/l. First-draw testing is important since young children might

normally take a drink of water in the middle of the night without proper
flushing. Infants might also get that lead-saturated first draw of water
in their liquid concentrated formula.

LEAD SOLDER

The random sample of homes selected for the South Huntington study
showed 67.3 percent of test sites have solder with a lead content between
55 and 65 percent. This tends to confirm that, during the past 20 years,
the predominant solder used in home plumbing in this area was lead solder.

Tin/Antimony and Tin/Silver Solder

Alternatives to lead solder include tin/antimony solder and tin/silver
solder. Recent wholesale costs per pound for these three solders in New
York State have been: tin/lead, $4.50; tin/antimony, $8.00; tin/silver,
$20.00. An estimate of one pound of solder per new home construction is
probably high.

The melting range of tin/silver and tin/antimony solder is slightly
higher than that for most tin/lead solder, 473°F for 95/5 tin/silver and
464°F for 95/5 tin/antimony, as opposed to 460°F for 40/60 tin/lead
solder. Tin/antimony solder has been tested for leaching in a prior U.S.
EPA/Seattle study which determined that the maximum leaching of antimony
was $3.7 \mu g/l$ after 27 hours of standing. The U.S. EPA/South Huntington
study found less than $4 \mu g/l$ leaching of antimony after overnight non-use
at a low pH.

Legislative Impacts of the Research

In New York State, the lead content of solder could be controlled in
one of three ways. The first is the State Building Code, which now permits
each of 931 towns and 553 villages to ban lead solder if the County Health
Department determines it a health hazard. The second method would be
through the State Plumbing Code, which has considered a ban on lead solder
for 18 months, but has taken no action. The third avenue is through state
legislation. Such legislation has been introduced, but no bill has yet
been reported out of committee. Based on the research conducted to date,
H2M has recommended both state legislation and local initiatives to limit
lead content in solder to 0.50 percent. In monitoring for compliance, we
have found a few tin/antimony solders that contain 0.30 percent and 0.35
percent lead. The Nassau County Health Department even found one with 0.50
percent lead. Based on our local findings, in order to assure compliance
without undue hardship on plumbers, we recommend that state and federal
legislation set the maximum lead content in solder at 0.50 percent.

On Long Island, 10 towns serving a population of 2,334,329 have
already moved to amend their plumbing codes in order to ban lead solder,
and some have instituted monitoring programs to test solder for its lead
content in order to verify that plumbers are adhering to the new code.

CONCLUSION

Leaching of metallic solders, particularly lead solder, appears to be affected by many factors:

- pH of water.

- Plumbing workmanship.

- Hardness of water.

- Time since last use of water.

- Age of solder.

- Percentage of lead in solder.

Where these conditions are conducive to lead leaching, large numbers of consumers are exposed to high-lead, first-draw water. If all evidence points to a need to ban lead solder -- why not now?

Part III

Impact of Plastic Pipe and Fittings
on Water Quality

IMPACT OF LEACHING BY PLASTIC PIPE, FITTINGS, AND JOINING COMPOUNDS

Thomas Podoll

SRI International
333 Ravenswood Avenue
Menlo Park, California 94025

The California Department of Housing and Community Development (DHCD) is proposing to expand the use of plastic plumbing pipe in residential construction. The principal change with regard to water quality is to allow polybutylene (PB) and chlorinated polyvinyl chloride (CPVC) pipe in hot and cold potable water supply systems inside dwellings. SRI International, under contract to the DHCD, prepared a document in March 1983 entitled "Environmental Review of Proposed Expanded Uses of Plastic Plumbing Pipe" (1). This environmental review provides an overview of existing information about the environmental impact of the new applications of plastic pipe and identifies areas where better information is needed. This presentation focuses on one part of this review, namely, the leaching of CPVC and PB plumbing.

OVERVIEW

The impact of leached chemicals can depend on either the instantaneous concentration of the chemicals in the water or the cumulative exposure to varying concentrations over time. Thus, we would ideally like to know concentrations of leachates as a detailed function of time after installation of plumbing systems. Figure 1 shows a decreasing concentration with time as the chemicals in the pipe are depleted or immobilized. Different chemicals would show more or less rapid declines over time after installation, depending on their properties and their initial distribution in the pipe.

The general pattern, however, is only a large-scale picture; many events disturb the smooth trend. Concentrations build up when water stands in the pipe for a period of time ("dwell time"), for example, overnight or during a vacation. When the system is flushed, concentrations drop to the background levels of the incoming water supply. The variation over a day or two might look something like that in Figure 2.

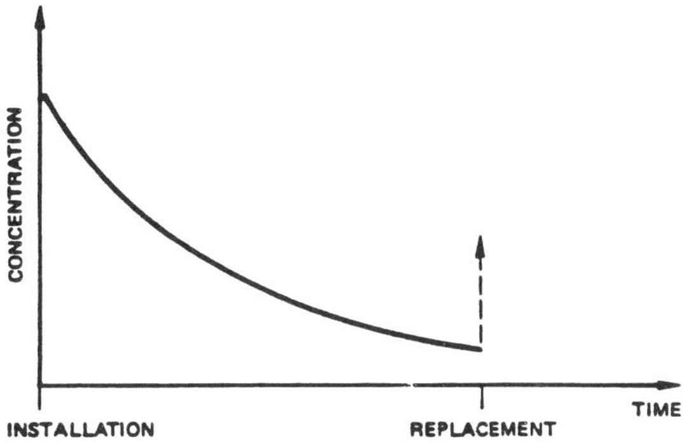

FIGURE 1   LONG-TERM PATTERN OF DECLINE IN CONCENTRATION

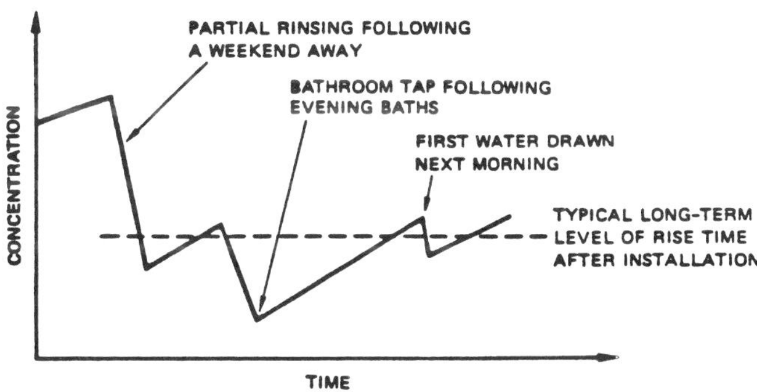

FIGURE 2   SHORT-TERM VARIATIONS IN CONCENTRATION

LEACHING MODELS

If water stands in a pipe, the concentration of a specific chemical in the water will be determined by the leaching rate, or flux, of the chemical integrated over time. The leaching rate, in turn, will be determined by the amount and distribution of the chemical in the pipe, the diffusion of the chemical in the pipe and in the water, the equilibrium partitioning of the chemical between the pipe and the water, and the diffusion of the leachate out of the pipe into the air.

The differential equations that describe mass transfer in the pipe and in the water are given by Fick's second law of diffusion. The solution of these equations depends on the initial and boundary conditions for the pipe/water system. Two relatively simple cases for pipe leaching appear to be reasonable. In the first case, a surface film of leachate covers the inner surface of the pipe and diffuses into an initially pure water phase. In the second case, the leachate is initially distributed uniformly in the pipe and diffuses into an initially pure water phase.

In Model 1 the leachate is concentrated at the pipe/water interface. The leachate equilibrates with the water relatively rapidly, and the amount of leachate in the water depends on the equilibrium partitioning of the leachate between the surface film and the water (described by H' in Figure 3) and on the number of equilibrium dwell periods that water has stood in the pipe. Figure 3 simply shows that as H' decreases (where the leachate becomes more soluble in water and less soluble in the pipe material), the concentration in the dwell water decreases more rapidly with the number of equilibrium dwell periods.

In Model 2 the leachate is initially uniformly distributed in the pipe. Equilibrium is not attained during a dwell period, and the amount of leachate in the water in a given dwell period will be diffusion controlled and over a long elapsed time will diminish with a square-root-of-time dependence. Figure 4 shows the concentration versus time profile for this model, assuming daily sampling and the concentration in units of the initial sample concentration.

Unfortunately, few data exist on pipe/water partitioning coefficients or plastic pipe diffusion coefficients and, because these coefficients are likely to be very sensitive to specific chemical/pipe interactions, they are difficult to predict. Although these models cannot be used to predict leaching from plastic pipes, they are useful for evaluating and understanding experimental data. This is particularly important for evaluating CPVC leaching data because of the complex composition of CPVC and the use of solvent cement to join the piping.

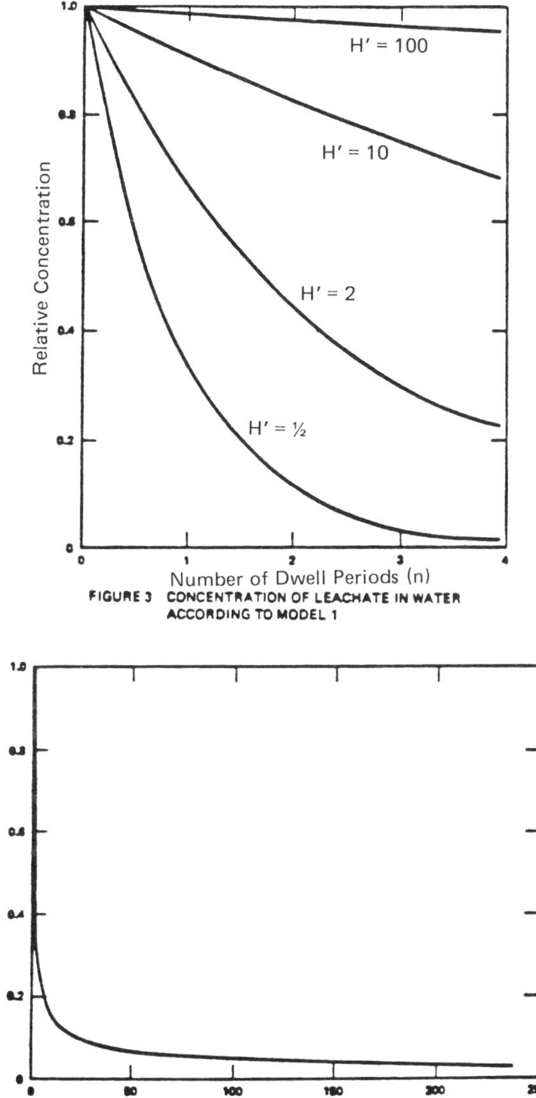

FIGURE 3   CONCENTRATION OF LEACHATE IN WATER
ACCORDING TO MODEL 1

FIGURE 4   CONCENTRATION VERSUS TIME PROFILE FOR MODEL 1

Assume daily sampling and concentration in units
of initial sample concentration.

PLASTIC PIPE COMPOSITION

The composition of plastic pipes should be known to determine possible leachates. Nonproprietary specifications supplied by manufacturers of plastic pipe and the National Sanitation Foundation (2) are summarized in Table 1. PB is composed primarily of the resin polymer and contains small amounts of pigments (such as titanium dioxide, carbon black, or talc) and a small amount of antioxidant (such as Irganox 1010). The composition of CPVC is more complex and requires a solvent cement for joining pipes. Potential health problems associated with toxic chemicals leached from joined CPVC pipes are accordingly more difficult to assess than potential water-quality problems with PB. Therefore, most of the remainder of this presentation focuses on CPVC leaching.

QUALITY ASSURANCE

There are several important reasons for closely examining the quality of plastic pipe leaching data:

1. The composition of pipe leachate water is likely to be a complex mixture (particularly for CPVC) of organic compounds in very dilute concentrations.

2. Many of the potential leachates of plastic pipe are commonly found in drinking water at measurable background concentrations.

3. Much of the work on plastic pipe leaching has been performed by parties that had institutional interest in the question of pipe water quality.

Excellent QA guidelines for plastic pipe leaching studies are detailed in the Interim Report by the National Sanitation Foundation entitled "Proposed Organohalide Leachate Testing Protocol for Plastic Piping."

CPVC LEACHING

The most important reports to date on CPVC leachability are those by James M. Montgomery (3) and Boettner et al. (4).

James M. Montgomery, Consulting Engineers, Inc., under contract to the California Department of Health Services, measured the water leachability of PVC and CPVC pipes. The testing protocol consisted of static, simulation, and kinetic tests. The static test was designed to estimate the concentrations of chemicals in water expected during initial occupancy. The kinetic test was designed to estimate the rate of leaching so that the concentration of the chemical in the water could be predicted as a function of time. The simulation test was designed to estimate leachability during typical home use.

TABLE 1

COMPONENTS OF PLASTIC PIPE SYSTEMS

PRODUCT	FUNCTION	PB	PERCENT	CPVC	PERCENT
Pipe and fittings	Resin	Poly(butene-1)	>98	Chlorinated PVC	>80
	Impact modifiers	None		ABS Methyl-methacrylate- butadiene-styrene- alpha-methyl styrene	>3 <15
	Stabilizers/ antioxidants	Irganox 1010	<0.5	Organotins	<3.5
	Lubricants	None		Oxidized polyethylene wax	<1.5
	Pigments/fillers UV stabilizers	Titanium dioxide Carbon black Talc	<2.0 <0.5 <2.0	Titanium dioxide Carbon black Others	<5.0 <0.05 <0.1
Solvent Cement	Solvents	None		Tetrahydrofuran Cyclohexanone 2-Butanone (methyl ethyl ketone) N,N-dimethyl formamide	80-90
	Resin/compound			CPVC or CPVC pipe compound	<20
	Pigments			Titanium dioxide Carbon black Others	<0.5
Primers	Solvents and pigments	None		Same as for solvent cement	

The major limitations of this study were a lack of replicate experiments and the consequent inability to quantify the statistical validity of the data. In spite of these and other limitations, the Montgomery report does give a semiquantitative picture of CPVC leaching.

Data compiled from the static and simulation tests indicate that the following organic chemicals were consistently found in significant concentrations in pipe water:

<u>CEMENT SOLVENTS</u>	<u>VOLATILE ORGANICS</u>
MEK	Dichloromethane
THF	Carbon tetrachloride
Cyclohexanone	Tetrachloroethene
DMF	Trichloroethene
	Toluene

The cement solvents were found in the parts-per-million range, but the amounts found appear to diminish rapidly, as expected, with rinsing. The volatile organics (except toluene) were typically found in the pipe water in the 1- to 10-ppb range for systems that approximated normal initial use of plumbing systems. Toluene was typically detected under 1 ppb. Chloroform may leach from the pipe; however, the data are inconclusive. Two of the tests found chloroform at levels no different from those in controls (kinetic and simulation), but it was found at levels marginally significantly different from those in controls in the static tests.

For the static tests, no correlation of concentration with dwell time or elapsed time was observed for the low-molecular-weight halogenated organics that were found in significant amounts. The cement solvents, however, show a definite decrease in concentration during the first few rinses of the cemented pipe. The absolute magnitude of these concentrations is probably unrealistic, even for initial use of a plumbing system, because of the low number of void volume rinses.

For the kinetic tests, no significant correlations between dwell time and concentration could be deduced for the volatile organics. Conversely, the concentrations of the cement solvents appeared to increase with increasing dwell time up to a limiting equilibrium value. Moreover, these concentrations fell significantly during the refill dwell kinetic experiments.

The work of Boettner et al. (1981) was initiated to develop an analytical method for detecting organotin stabilizers in water. In the process, useful information on CPVC leachability was obtained.

This report confirmed the findings of the Montgomery report on the leachability of cement solvents. High initial concentrations of these solvents were found in the pipe water and these concentrations decreased with elapsed time. Boettner et al. also detected the presence of organotin leachates in CPVC pipe water. The concentrations of these leachates were initially in the 0.5- to 3-ppb range for 1-day dwell periods, but fell rapidly to less than 0.1 ppb per day after 3 weeks elapsed time.

It is interesting to note that the concentrations of the organotin leachates appeared to diminish in a biphasic manner with time; that is, the concentrations diminished, then increased, then diminished again. This biphasic behavior suggests two types of leaching in cement-joined pipes. The initial leaching of organotins occurs from areas of the pipe where solvent cement has not been applied. This leaching diminishes rapidly as the surface concentration is depleted. In the areas where solvent cement has been applied, there is no organotin present initially, but within a few days, organotin species begin to diffuse through the cement/pipe surface layer into the water, and the leachate concentration correspondingly rises.

These data suggest a possible explanation for a lack of correlation between leachate concentrations of noncement solvents and time that was described in the 1980 Montgomery report. If solvent cements were applied in differing thicknesses to the joints of the test systems, it would be expected that the leaching of pipe constituents through the cement to the water would be retarded in varying degrees. Thus, while the leaching rate in one part of the system may have been diminishing, the leaching rate in another part of the system may have been increasing. For a plumbing system with several joints, the variability of the leaching rate with elapsed time over a period of several weeks may well have been sporadic, as observed.

SUMMARY OF PLASTIC PIPE LEACHING DATA

CPVC leaching data indicate that:

- Pipe cement solvents (MEK, THF, cyclohexanone, and DMF) are initially leached at very high concentrations, but these concentrations diminish rapidly with repeated flushing.

- Low-molecular-weight chlorinated organics (dichloromethane, carbon tetrachloride, tetrachloroethene, and trichloroethene) are leached in the low (1-10) ppb concentration range during initial dwell periods. The lack of correlation of the concentrations of these leachates with dwell time or elapsed time is probably caused by the relative uncertainty of the measured values (which were typically within a factor of 5 of control values) and by the cumulative effect of leaching of these organics at different rates from areas covered or not covered with solvent cement.

- Chloroform is a possible leachate. More work is required on chloroform leaching and on the leaching of chemicals other than the dominant solvents from the solvent cements.

POLYBUTYLENE LEACHING

The leachability studies of PB were generally of lower quality than those of CPVC. However, a study by Shell (5) indicated the presence of Irganox 1010 derivatives in leachate waters. Irganox 1010 is an antioxidant that is present in PB at less than 0.5%.

REFERENCES

1.   SRI International. 1983. "Environmental Review of Proposed Expanded Uses of Plastic Plumbing Pipe." Prepared for California Department of Housing and Community Development, Menlo Park, California.

2.   National Sanitation Foundation. 1983. "Proposed Organohalide Leachate Testing Protocol for Plastic Piping," Interim Report.

3.   James M. Montgomery, Consulting Engineers, Inc. 1980. "Solvent Leaching from Potable Water Plastic Pipes," Final Report. Prepared for the Hazard Alert System, California Department of Health Services/Department of Industrial Relations, James M. Montgomery, Consulting Engineers, Inc., Pasadena, California.

4.   Boettner, E.A., G.L. Ball, Z. Hollingsworth, and R. Aguino. 1981. "Organic and Organotin Compounds Leached from PVC and CPVC Pipe," U.S. Environmental Protection Agency, Cincinnati, Ohio.

5.   Shell Chemical Company. 1982. "Analysis of Water Leachates of Polybutylene (PB) Pipe Supporting the Conclusion that the Use of PB Pipe for Domestic Water Service Is Without Measurable Hazard," Shell Chemical Company, Houston, Texas.

NSF STANDARD AND CERTIFICATION PROGRAM
FOR PLASTICS PIPE: STANDARD 14

Nina I. McClelland

President and Chief Executive Officer
National Sanitation Foundation
3475 Plymouth Road
Ann Arbor, Michigan 48105

The National Sanitation Foundation (NSF) is a private, not-for-profit corporation, chartered in 1944 under Michigan law. Its mission is to develop and administer programs relating to public health and the environment in areas of services, research, and education. Through its headquarters and laboratories in Ann Arbor, Michigan, a wastewater equipment testing facility in Chelsea, Michigan, and regional offices in or near Los Angeles (Upland) and Sacramento (Davis), California; Ann Arbor, Michigan; Philadelphia (Chalfont), Pennsylvania; Atlanta, Georgia; and Dover, England, NSF's programs and services reach throughout the United States and into 25 foreign countries.

For operational purposes, programs are grouped into Listing, Certification, and Assessment Services. Listing Services programs are those for which there is an NSF standard, authorization to display the appropriate seal or logo, and a published product Listing. Certification Services programs involve standards other than NSF's official regulations, or specifications. They include authorization to display the certification mark, and a published product Registry. All other programs and special studies are provided through Assessment Services. A controlled use report is provided for special studies; the identity mark is the company logo. Each of these programs is available for domestic and international application.

A highly qualified professional and technical staff — engineers, chemists, microbiologists, and environmental scientists — is retained by NSF in state-of-the-art facilities. Consulting faculty from the University of Michigan are retained for toxicological, radiological, and other expertise, as required. Recent capital acquisitions focus on laboratory and information processing resources intended to optimize quality and timeliness of all operations. A \$3.25 million laboratory expansion program (current) attests to the high performance and growth objectives to which the Trustees and staff are fully committed.

Plastics piping system components have been tested at NSF since 1955, first in a comprehensive special study of leachate and toxicological testing (1), then under the provisions of Standard 14, Plastic Piping System Components and Related Materials. This Standard includes products for potable water; drain, waste, and vent; and other plumbing systems applications.

Ingredients used in products listed for potable water must comply with rigid acceptance and qualification procedures prior to listing. These requirements are summarized in Figure 1.

With an application for listing, a manufacturer provides complete, detailed chemical identity of all ingredients in the formulation. This information is reviewed by staff, and held under terms of confidential disclosure. For ingredients generally regarded by the Food and Drug Administration (FDA) as safe, or sanctioned previously for food contact, e.g., listed in Title 21, Code of Federal Regulations (21CFR), no further toxicological testing is usually required. For nonsanctioned ingredients, Ames and animal feeding studies must be provided. The animal data must include established no-effect, effect, and intermediate levels of the proposed ingredient, using protocols and a testing laboratory accepted by NSF in advance of the testing.

Chemical leachate testing is required for all proposed new ingredients, using pipe and compound formulated to contain the ingredient at two times the level proposed for maximum use. Chemical leachate (extraction) testing for inorganic contaminants is performed consistent with the testing protocol in Standard 14; i.e., exposure to "formulated water" at pH 5.0 $\pm$ 0.2 at 37 degrees (°C) for periods of 24, 24, and 72 hours. (Cold-water end-use applications are assumed; for hot-water applications, exposures are 1 hr at 82° C, 1 hr at 82° C, and 72 hr at 37° C.) Fresh water is added following each of the 24-hour exposures, providing data for three separate extractions. The first exposure is assumed to simulate worst- case end use, where the user would ingest the first, very aggressive water drawn from a new installation; the third exposure represents water which could be ingested as the first draw of water left standing over a weekend. The protocol for organic chemical leachate testing is similar except that pH 8.0 $\pm$ 0.15 is used for the exposure.

The chemical parameters monitored and their maximum permissible limits (MPLs) are shown in Figure 2. (Currently, five volatile organic compounds proposed by EPA for regulation are also being monitored to acquire a data base. They are CCl_4; TCE; PCE; 1,2-dichloroethane; and 1,1,1-trichloroethane.) For proposed new ingredients, the MPL for the first exposure (MPL-1) is set at ten times the established, third-exposure MPL (MPL-3). MPL-3 is equivalent to maximum contaminant levels (MCLs) in the National Interim Primary Drinking Water Regulations (NIPDWR) for all regulated chemicals included in Standard 14. By policy, all chemicals regulated by NIPDWR are included in Standard 14 when pertinent to the listed products.

Acceptance[1] (New or Generically Similar Ingredient)	Qualification[1] (New Product or Change in Formulation)	Monitoring[2]
Disclosure statement with complete chemical identity Product and compound tested at 2 X maximum recommended use level Chemical leachate testing[3] 1st exposure $\leq$10 X MPL[4] 3rd exposure $\leq$MPL[4]	Product and compound tested at maximum use level Chemical leachate testing[3] 1st exposure $\leq$10 X MPL[4] 3rd exposure $\leq$MPL[4] Performance testing HDS ASTM[4]	Chemical extraction testing[3] 3rd exposure $\leq$MPL[4] Performance testing HDS ASTM[4]
FDA/21CFR If no, Ames test and 90-day feeding study at effect, no effect, and intermediate levels		

[1]Samples submitted by applicant.
[2]Samples collected by regional staff.
[3]RVCM $\leq$MPL is also required for PVC and CPVC pipe and fittings.
[4]Refers to all parameters in Standard 14.

Figure 1.   Summary of requirements for products listed under Standard 14.

Parameter	MPL-3 (ppm)
Antimony	0.05
Arsenic	0.05
Barium	1.0
Cadmium	0.01
Chromium	0.05
Lead	0.05
Mercury	0.002
Phenolic Substances	0.05
RVCM (in finished product)	10.0
Selenium	0.01
Tin	0.05
TTHMs	0.01

Figure 2.   Chemical parameters in Standard 14.

A special test protocol has been developed and validated for organics leachate testing. Nineteen samples of polyvinylchloride (PVC) and chlorinated polyvinylchloride (CPVC) were used for validation testing. To date, none of the data suggests that the listed products contribute significant levels of trihalomethanes to contacted water.

To confirm compatibility with other ingredients in a product recipe, an accepted new ingredient must also be "qualified" by additional chemical leachate testing. Products formulated by the ingredient user (pipe, fittings, or appurtenance manufacturer) at the maximum proposed use level are tested in accordance with the procedure described for acceptance testing.

Following acceptance of an ingredient and qualification of product, listing is authorized. The appropriate logo is then displayed on the product; i.e., NSF-pw, -wc, for potable water/well casing applications; NSF-dwv, -tubular, for drain, waste, and vent/continuous waste systems; NSF-cw, for corrosive wastes; NSF-sewer, for sewer main applications.

Standard 14, adopted in 1965, is revised and updated on a continuing basis. In 1978, a requirement for residual vinyl chloride monomer (RVCM) was added. By modeling and leachate testing, it was established that levels to 10 parts per million (ppm) RVCM in the wall of pipe or fittings would not leach detectable levels of the monomer to water exposed to product (where "detection" is 2 ppb). Results of RVCM monitoring experience show that none of the 460 samples tested in 1982 or 472 samples tested in 1983 exceeded the established MPL of 10 ppm.

From the outset, standards development at NSF has included representatives at all levels of government, the affected industry, and users of the subject products or services. Typically, the request for a standard may originate with any of the three sectors. Interest and commitment are established in an exploratory meeting. A small task committee is appointed to draft the document. The draft is reviewed and revised or accepted by the Joint Committee (JC), where each sector -- regulatory, industry, and user -- has voting representation. The Joint Committee proposes the standard to the Council of Public Health Consultants, a group of 36 public health professionals with no industry representation. The Council recommends a standard to the Board of Trustees, where final adoption occurs.

All NSF standards have a requirement for periodic review at intervals not to exceed five years. The formal, ongoing review and revision or reaffirmation process is accomplished through established Joint Committee policies. All voting is in compliance with procedures described in the Office of Management and Budget (OMB) Circular A119 (2).

Three annual inspections are required for all plastics production facilities included in the listing. These are minimums and apply to foreign and domestic sites. Nonconformance and other problems result in

additional inspections. At the plant, products in production and inventory
are checked to verify consistency with previous design and testing
records. Use of accepted ingredients or compounds and in-plant quality
assurance are verified; and selected evaluations may be performed (e.g.,
dimensioning products). Seven hundred and eleven (711) inspections are
projected for 1984. Violations identified either in testing or inspection
result in various actions, ranging from corrective measures to delisting.

There are three general categories of regulation — official,
self-regulation, and third-party. Third-party programs have established
credibility world-wide. They effectively and appropriately place cost
burdens with the private sector. Their objectivity and long-term success
are a matter of record and a source of pride! We look forward to expanded
opportunities with plastics plumbing systems components and related
products through voluntary consensus standards and third-party
certification programs like NSF Standard 14.

REFERENCES

1.   Tiedeman, W. and N.A. Milone. 1955. A Study of Plastic Pipe for
     Potable Water Supplies, NSF, June 1955.

2.   Office of Management and Budget. 1982. OMB Circular A119, Federal
     Participation in the Development and Use of Voluntary Standards, Vol.
     47, November 1, 1982.

EVALUATION OF THE PERMEATION OF ORGANIC SOLVENTS THROUGH GASKETED JOINTED AND UNJOINTED POLY(VINYL CHLORIDE), ASBESTOS CEMENT, AND DUCTILE IRON WATER PIPES

James P. Pfau

Group Leader, Coating Science, Polymer Science and Technology Section
Battelle Columbus Laboratories
505 King Avenue
Columbus, Ohio 43201

Donald Goodman

Manager, Polymer R&D
Tenneco Polymers, Inc.
R.D. 6, Box 177
Flemington, New Jersey 08822

SUMMARY*

Phase one of the Vinyl Institute's research program, "Evaluation of the Permeation of Organic Solvents through Gasketed Jointed and Unjointed Poly(vinyl chloride), Asbestos Cement, and Ductile Iron Water Pipes," has been completed. Phase one involved the exposure of commercially available 4-inch water pipe joints to three neat organic solvent environments (toluene, hexane, and 1,1,1-trichloroethane) for a period of 6 weeks. Each pipe specimen contained demineralized water under a pressure of 40 pounds per square inch gauge (psig) and throughout the 6-week exposure period, samples of water were removed from each pipe and analyzed for traces of the appropriate organic liquid.

The objective of this program was to study when and how permeation of selected solvents occurs through the jointed and unjointed pipes and to develop a hypothesis for the mechanism of permeation. The tests results indicate that the mechanism for fastest breakthrough (permeation) is through the gasketing material. The unjointed pipe showed slower permeation or no permeation compared to the jointed pipe. When permeation

---

*Copies of original work are available on request to: Dr. Roy T. Gottesman, Executive Director, Vinyl Institute, 355 Lexington Avenue, New York, New York 10017.

82

did occur with the jointed pipe, it happened much earlier than with the straight unjointed pipe. Another mechanism for breakthrough is based on the composition of the pipe. The moreporous asbestos/cement unjointed pipe showed early permeation with toluene and 1,1,1-trichloroethane, and the unjointed PVC pipe swelled to permeation with toluene after day 37 of the 42-day test.

The initial work was performed under "worst case model" conditions. The breakthrough of neat solvents in the jointed test specimens indicates that the joints of all three materials are susceptible to permeation. The second phase of the program, expsoure to aqueous solutions of the same organic solvents, appears to be warranted in order to determine permeation under more realistic conditions. Consideration should be given in characterizing the rubber compounds used in the gaskets. Differences in the gaskets used with each type of pipe may explain the apparent differences in permeation, particularly with regard to hexane.

CONCLUSIONS

1. Permeation of neat solvents is related to type of solvent, type of pipe, whether the pipe is jointed or unjointed, and time of exposure.

2. For all the organic solvent/pipe combinations investigated, the jointed pipes showed organic solvent breakthrough (permeation). The sole exception was the ductile iron/hexane, although the jointed ductile iron pipe demonstrated permeation with toluene and 1,1,1-trichloroethane.

3. The unjointed pipe showed slower permeation or no permeation compared to the jointed pipe. Also, when permeation did occur with the jointed pipe it happened earlier than with the straight unjointed pipe.

4. The permeation data indicate that the mechanism for fastest breakthrough is through the gasketing material. The gasketing material used to seal the jointed pipes is more susceptible to organic solvent (toluene, hexane, and 1,1,1-trichloroethane) permeation than is the pipe material.

5. Another mechanism for permeation is based on the composition of the pipe. The more porous asbestos/cement unjointed pipe showed early permeation by toluene and 1,1,1-trichloroethane. The unjointed PVC slowly softened to point of permeation with toluene between sampling days 37 and 41, and the unjointed iron pipe did not show permeation.

6. The order of aggressiveness (i.e., attack on the pipes and gaskets) for the organic solvents from greatest to least is toluene > 1,1,1-trichloroethane > hexane.

7.   Toluene permeated all jointed pipe sections. Toluene did not
     permeate the ductile iron unjointed pipe, and the PVC unjointed
     pipe showed breakthrough after day 37 of the 42-day test.

8.   1,1,1-Trichloroethane permeated all jointed pipe sections. It
     did not permeate the PVC straight pipe or the iron straight
     pipe. The asbestos straight pipe showed permeation at about 2
     weeks.

9.   Hexane permeated PVC jointed pipe and asbestos/cement jointed
     pipe. Iron jointed pipe and all straight pipes showed no
     permeation. The presence of a dark oily residue in the jointed
     PVC pipe substantiates the conclusion that the gasketing material
     has been attacked. The absence of permeation with jointed iron
     pipe suggests that this gasket is not solvated by hexane, whereas
     it was attacked by toluene and 1,1,1-trichlorethane in parallel
     tests.

10. Entry of organic solvent into the demineralized water against 40
     psig internal pressure is corroborating evidence that the
     mechanism is permeation through gasketing material or pipe and
     not hydraulic flow. Constant internal pressure does not exclude
     contamination by permeation.

Part IV

Solutions to Plumbing-Related Water Quality Problems

TREATMENT OR WATER QUALITY ADJUSTMENT TO ATTAIN MCLs IN METALLIC POTABLE WATER PLUMBING SYSTEMS

Michael R. Schock

Associate Chemist, Aquatic Chemistry Section
Illinois State Water Survey
6050 E. Springfield
Box 5050, Station A
Champaign, Illinois 61820

INTRODUCTION

Virtually all of the common household or building plumbing materials — copper tubing, galvanized steel, lead pipe, brass fittings, and tin/lead solder — will oxidize and dissolve to some extent in potable waters. The byproducts of these corrosion reactions can be of concern because of aesthetic or toxicological reasons; hence, the development (1,2) of maximum contaminant levels (MCL) or of secondary (2) maximum contaminant levels (SMCL) for lead, cadmium, copper, zinc, and iron. Additionally, corrosion is economically costly, because of decreased service lifetimes of piping materials, loss of water during distribution, and decreases of hydraulic efficiency.

Traditionally, corrosion control measures taken by water utilities have been aimed at preventing tuberculation and perforation of cast-iron distribution mains, eliminating "red water" complaints, and protecting dissolution of cement-mortar-lined or asbestos-cement pipe. Treatment processes designed to alleviate some or all of these distribution system problems may or may not effectively limit the corrosion of the different types of metallic materials used in the service lines and interior systems of buildings and dwellings.

The purpose of this paper is to examine the impact of several common treatment approaches on the formation of corrosion byproducts from common building plumbing materials. The treatment approaches covered in this paper will be: calcium carbonate saturation, pH adjustment, pH plus carbonate adjustment, orthophosphate addition, polyphosphate addition, and silicate addition. First, however, a background will be developed to aid in the understanding of the feasibility of different corrosion control approaches.

SOLUBILITY AND MASS-TRANSPORT PROCESSES

Unlike most distribution system mains, household and building systems frequently have large sections of piping in which the water can be stagnant for hours. During these long periods of standing, corrosion reactions can continue to add byproducts to the water until a state of solubility equilibrium is nearly or actually attained. Thus, the metal levels observed in the water are initially governed by the mass transport of oxidizing or solubilizing agents to the pipe surface, and of the corrosion products away from the surface. The upper-limiting case, from a dissolved byproduct standpoint, is that ultimately provided by the solubility equilibria of the surface pipe material in the water. Physical factors, such as nonadherence of the corrosion product film or the dislodging of scales by normal turbulent flow, can add particulate metal species to the water, increasing the observed "total" metal concentration. Also, corrosion reactions can continue, even if the solubility equilibrium is attained.

Clearly, there are numerous tradeoffs in corrosion control programs. For instance, the minimization of metal levels in order to attain MCLs may not result in the formation of an effective barrier to continuing metal attack, shortening the lifetime of the plumbing material. It may merely convert the metal into an adherent corrosion product while surface attack and/or pitting continues. Alternatively, an absence of bulky corrosion products and tubercles on iron lines brought about by treatment chemicals, such as polyphosphates, might result in increased levels of soluble iron through complexation.

In general, the strategy of corrosion control through water quality adjustment involves interference with at least one link of the chain of interconnected anodic and cathodic corrosion cell reactions (4). Ordinarily, this consists of the formation of a "barrier" of a "passivating film" that both limits the transport of the metallic species into solution (mostly through solubility limitation), and limits the diffusion of the oxidizing agents (normally dissolved oxygen or chlorine species) to the pipe surface. These complicated couplings of redox and solubility reactions, plus the differences in relative rates of the reactions, prevent simple relationships between corrosion rates determined from pipe coupon or insert data and solubility levels from being developed.

An example of the role of mass transport in governing the metal level in nonequilibrium systems has been elegantly developed by Kuch and Wagner (5) for lead. The approach is generally suitable for other types of pipe, as well. These authors show that the level of lead in a water undergoing turbulent flow can be predicted by knowing the lengths and diameters of the lead pipes, the observed (or computed) concentration of lead at solubility equilibrium, the volume flow rate of the water, the water temperature, and the diffusion coefficient of the dissolved species. For water under stagnating conditions, the parameters that need to be known to compute the lead level after a given stagnation time are: the diameter of the pipe, the diffusion coefficient (approximated by that for Pb^{2+}), the maximum equilibrium lead concentration, and the lead concentration during flushing (computed from the previous model). In hard waters, the mass transfer coefficient may also need to be included.

FORMATION OF PASSIVATING FILMS

Because the _in situ_ formation of tight, adherent films that are impervious to ionic or oxidant migration is virtually impossible, a discussion of the solids that can precipitate in pipes and produce a limit for metal solubility is in order. Only common building plumbing materials — lead, copper, new galvanized steel, and brass — will be considered here.

Table 1 gives a list of solid corrosion products that have been found that are likely to be significant in controlling the solubility of the different pipe or coating metals. Each metal behaves uniquely, and these solids are largely responsible for determining whether or not a corrosion control technique will work. Other deposits have been reported, but systematic and comprehensive surveys in many water qualities at different temperatures have not been done. Solids containing sulfate and chloride are often common in saline waters, waters with a high ionic strength, or in the presence of pits in copper tubing (6). Other deposits may also locally form, but they may not act as solubility controls.

Detailed research has not been performed to identify the films of corrosion products or inhibitor precipitates formed on tin/lead solder or brass fittings under different water quality conditions. Until that time, we must assume formation of the same products as would be present on pipes of similar material.

The relationship of passivating film formation to water quality adjustment and inhibitor dosage will be covered for galvanized steel (zinc), copper, and lead in the next section. To assist in predicting orthophosphate inhibitor effectiveness and the effect of pH and alkalinity (through the inorganic carbonate content), solubility calculations were performed in a similar manner to those reported previously for lead (7-10) and zinc (10-11). A new computer program was created for calculation of copper (II) solubility. The equilibrium constants used are reported in Tables 2 and 3. Considerable uncertainty exists for much of the thermodynamic data, and all computations were restricted to 25°C because temperature effects on the equilibrium constants are largely unexplored. An ionic strength of 0.0005 moles per liter (mol/L) was assumed as a general approximation. The pH interval of 7 to 11 was considered, as well as total inorganic carbonate (TIC) concentrations of from 5 to 500 milligrams calcium carbonate per liter (mg $CaCO_3$/L) (0.05-5.0 mmole C/L). Only orthophosphate, hydroxide, and carbonate solids with a documented history of occurrence in plumbing systems were included; hence,

**Table 1. Scale Minerals That May Be Important in
Limiting the Solubility of the Indicated Metals and Metallic
Materials in Most Potable Water Systems**

<u>Galvanized Steel Pipe</u>

Basic Zinc Carbonate (hydrozincite, $Zn_5(OH)_6(CO_3)_2$)
Zinc Hydroxide, $Zn(OH)_2$, α-, β-, δ-, and ϵ- forms
α-Hopeite, $Zn_3(PO_4)_2 \cdot 4H_2O$
Zinc Oxide (Franklinite), ZnO or $ZnFe_2O_4$
Hemimorphite, $Zn_4Si_2O_7(OH)_2 \cdot H_2O$ (hot water systems)

<u>Copper Pipe</u>

Cuprite, Cu_2O
Tenorite, CuO
Basic Copper Carbonate (Malachite), $Cu_2(OH)_2CO_3$
$Cu_3(PO_4)_2 \cdot 4H_2O$
Brochantite, $Cu_4(OH)_6SO_4$
Atacamite, $Cu_2(OH)_2Cl$

<u>Lead Pipe</u>

Basic Lead Carbonate (hydrocerussite), $Pb_3(CO_3)_2(OH)_2$
Lead Carbonate (cerussite), $PbCO_3$
Hydroxypyromorphite, $Pb_5(PO_4)_3OH$
Plattnerite, PbO_2

Table 2. Reactions and Log Equilibrium Constants (β) at 25°C
and I = 0 mol/L Species Considered in Solubility Diagram Construction
M = Pb, Cu or Zn. All constants are for I = 0 and 25°C

			log β		
			Pb	Zn	Cu
$M^{2+} + H_2O$	$\leftrightarrow$	$MOH^+ + H^+$	-7.23	-8.96	-7.93
$M^{2+} + 2H_2O$	$\leftrightarrow$	$M(OH)_2^{\circ} + 2H^+$	-16.93	-16.9	-16.23
$M^{2+} + 3H_2O$	$\leftrightarrow$	$M(OH)_3^- + 3H^+$	-28.10	-28.4	-26.9
$M^{2+} + 4H_2O$	$\leftrightarrow$	$M(OH)_4^{2-} + 4H^+$	-39.7	-41.2	-39.6
$2M^{2+} + H_2O$	$\leftrightarrow$	$M_2OH^{3+} + H^+$	-6.36	-9.0	--
$2M^{2+} + 2H_2O$	$\leftrightarrow$	$M_2(OH)_2^{2+} + 2H^+$	--	--	-11.21
$3M^{2+} + 4H_2O$	$\leftrightarrow$	$M_3(OH)_4^{2+} + 4H^+$	-23.88	--	-22.05
$4M^{2+} + 4H_2O$	$\leftrightarrow$	$M_4(OH)_4^{4+} + 4H$	-20.88	--	--
$6M^{2+} + 8H_2O$	$\leftrightarrow$	$M_6(OH)_8^{4+} + 8H^+$	-43.61	--	--
$M^{2+} + CO_3^{2-} + H_2O$	$\leftrightarrow$	$MHCO_3^+$	--	11.73	12.41
$M^{2+} + CO_3^{2-}$	$\leftrightarrow$	MCO_3°	7.1	5.2	6.8
$M^{2+} + 2CO_3^{2-}$	$\leftrightarrow$	$M(CO_3)_2^{2-}$	10.3	7.5	9.8
$M^{2+} + PO_4^{3-} + H^+$	$\leftrightarrow$	$MHPO_4^{\circ}$	15.45	15.66	16.36
$M^{2+} + PO_4^{3-} + 2H^+$	$\leftrightarrow$	$MH_2PO_4^+$	21.1	21.16	21.33

Table 3. Equilibrium Constants at 25°C and I = 0 mol/L for the
Solids Considered in the Exploratory Solubility Calculations

			log K
$Pb(OH)_2(s) + 2H^+$	$\leftrightarrow$	$Pb^{2+} + 2H_2O$	13.07
$PbCO_3(s)$	$\leftrightarrow$	$Pb^{2+} + CO_3^{2-}$	-13.13
$Pb_3(CO_3)_2(OH)_2(s) + 2H^+$	$\leftrightarrow$	$3Pb^{2+} + 2CO_3^{2-} + 2H_2O$	-18.0
$Pb_5(PO_4)_3OH(s) + H^+$	$\leftrightarrow$	$5Pb^{2+} + 3PO_4^{3-} + H_2O$	-62.8
$CuO(s) + 2H^+$	$\leftrightarrow$	$Cu^{2+} + H_2O$	-7.62
$Cu(OH)_2(s) + 2H^+$	$\leftrightarrow$	$Cu^{2+} + 2H_2O$	-8.64
$Cu_2(OH)_2CO_3(s) + 2H^+$	$\leftrightarrow$	$2Cu^{2+} + CO_3^{2-} + 2H_2O$	-5.18
$CuCO_3(s)$	$\leftrightarrow$	$Cu^{2+} + CO_3^{2-}$	9.63
$Cu_3(PO_4)_2 \cdot 2H_2O(s)$	$\leftrightarrow$	$3Cu^{2+} + 2PO_4^{3-} + 2H_2O$	-35.12
$\varepsilon-Zn(OH)_2(s) + 2H^+$	$\leftrightarrow$	$Zn^{2+} + 2H_2O$	11.54
$ZnCO_3(s)$	$\leftrightarrow$	$Zn^{2+} + CO_3^{2-}$	-10.00
$Zn_5(OH)_6(CO_3)_2(s) + 6H^+$	$\leftrightarrow$	$5Zn^{2+} + 2CO_3^{2-} + 6H_2O$	9.65
$Zn_3(PO_4)_2 \cdot 4H_2O(s)$	$\leftrightarrow$	$3Zn^{2+} + 2PO_4^{3-} + 4H_2O$	-35.3

the absence of solids such as $Pb_3(PO_4)_2$ and $Pb_3(OH)_5Cl$. An exception was
$Cu_3(PO_4)_2 \cdot 2H_2O$, which was the only copper orthophosphate for which data
was available.

The results of some of the computations are presented in Figures 1
through 5, which will help show trends in solubility behavior of three of
the trace metals of interest. Even though computations went up to 2 mg
PO_4/L, the diagrams are shown only for an orthophosphate concentration of
0.5 mg PO_4/L, which is a common dosage. For copper (II), no
precipitation of orthophosphate was predicted. These diagrams show
"contour lines" for the metal solubility, as both pH and TIC are varied. A
three-dimensional version of Figure 1 for lead (II) is shown in Figure 6,
for comparison. The units of "log mg/L" were chosen to produce smooth
contour line intervals and good detail in the diagrams. Reference will be
made to these figures in the following section.

TREATMENT APPROACHES

CALCIUM CARBONATE SATURATION

This is the approach traditionally followed by utilities with hard and
moderately hard water supplies. Also, many utilities who use lime
softening for clarification also use this strategy. Monitoring the calcium
carbonate saturation state was also included as one of the requirements of
the U.S. EPA "Corrosivity" regulation (1). Many expositions on the
rationale for $CaCO_3$ saturation are available, but the primary intent can
be summarized by saying that the objective has been to prevent
tuberculation and "red water" in iron distribution mains by covering the
pipe with a thin and adherent layer of calcium carbonate. However,
research by Stumm in the 1950s and 1960s (12, 13), and more recently by
primarily German scientists (14, 15), indicates that the effective
corrosion control mechanism involves the formation of a dense siderite
($FeCO_3$) scale layer, and is also related to the buffer intensity of the
water and the Fe(II)/Fe(III) ratios in the pipe deposits.

Numerous indices are available for the estimation of the calcium
carbonate saturation state, and they have been thoroughly reviewed
elsewhere (16-18). The concept of using a calcium carbonate film to "seal"
the surface of a copper, lead, or galvanized pipe in household or building
systems in order to prevent metal leaching has been widely assumed to have
been validated. However, virtually no research has been published to
define the water quality conditions (i.e., pH, temperature, calcium, and
carbonate concentrations) or the time necessary to form the film, or to
examine the metal leaching behavior while film formation is in progress.
For zinc and copper, Figures 3 and 5 indicate that concentrations below the
SMCLs should remain easily attainable at pH values above 7.5 for most
conceivable TIC levels, in the absence of complete surface coverage by the
calcium carbonate deposit. Obviously, the thickness and homogeneity of the

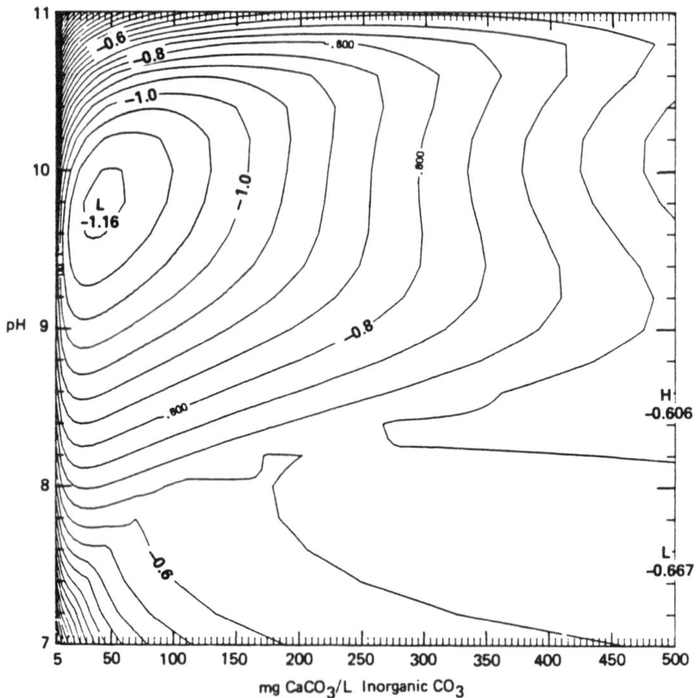

Figure 1.   Contour diagram of lead (II) solubility in the system Lead (II)-Water-Carbonate at 25°C and an ionic strength of 0.005 mol/L.   The MCL is -1.30 (0.05 mg Pb/L).   The points marked "L" and "H" indicate local low and high points.   The contour interval is 0.05 log units.

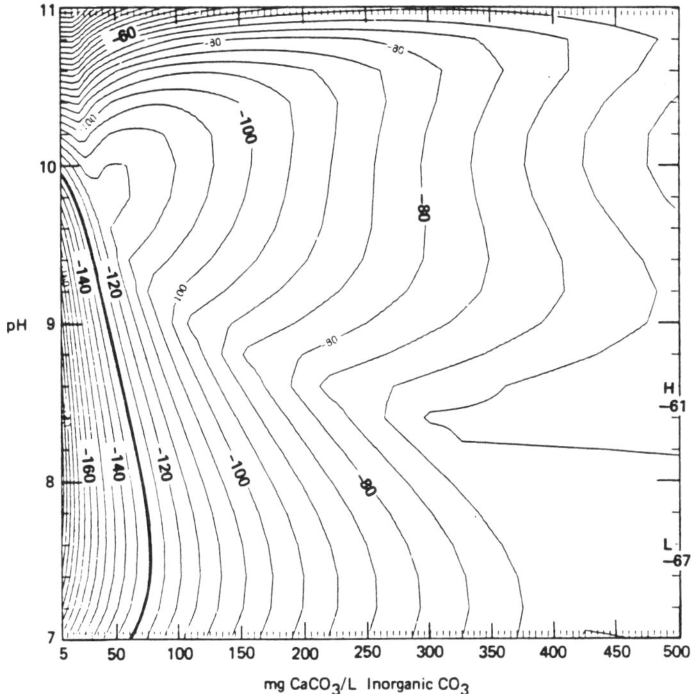

Figure 2. Contour diagram of lead solubility in the presence
of 0.5 mg PO$_4$/L at I = 0.005 mol/L and 25°C. The MCL =
-130 on this scale, which is: Log (mg Pb/L) x 100. The
contour interval is 0.05 log units.

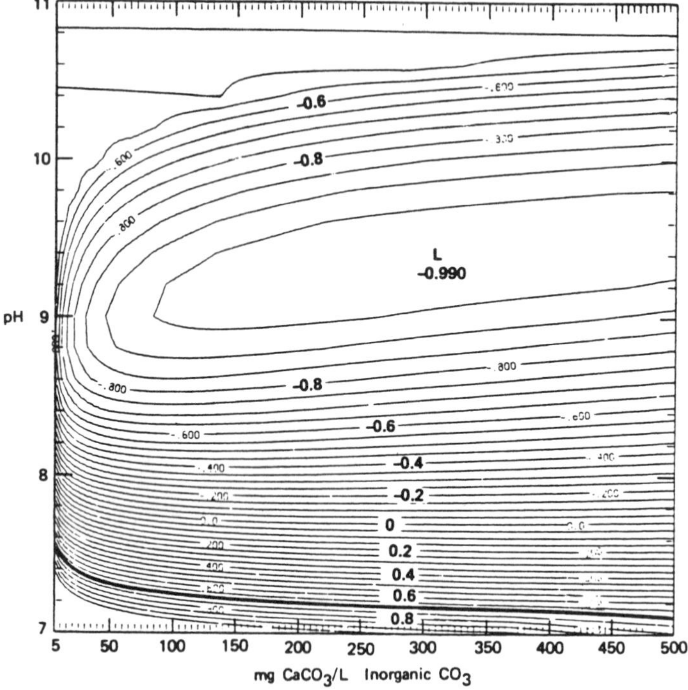

Figure 3. Contour diagram of zinc solubility in the system
Zinc-Water-Carbonate at 25°C and I = 0.005 mol/L. The
SMCL is 0.699 (5 mg Zn/L), and the contour interval is
0.05 log units.

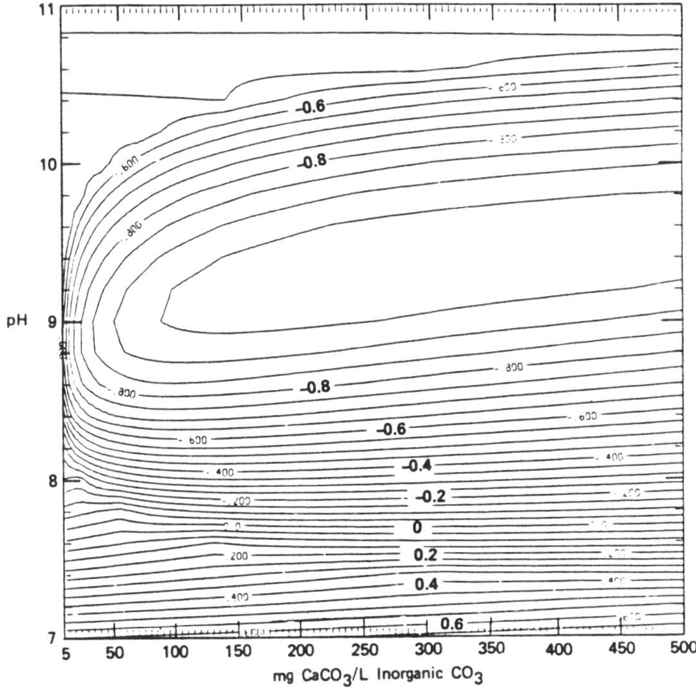

Figure 4. Contour diagram of zinc solubility in the presence of 0.5 mg PO4/L at I = 0.005 mol/L and 25°C. All points lie below the 5-mg Zn/L SMCL (0.699 in diagram units). The contour interval is 0.05 log units. Lines were removed in the lower left to avoid crowding.

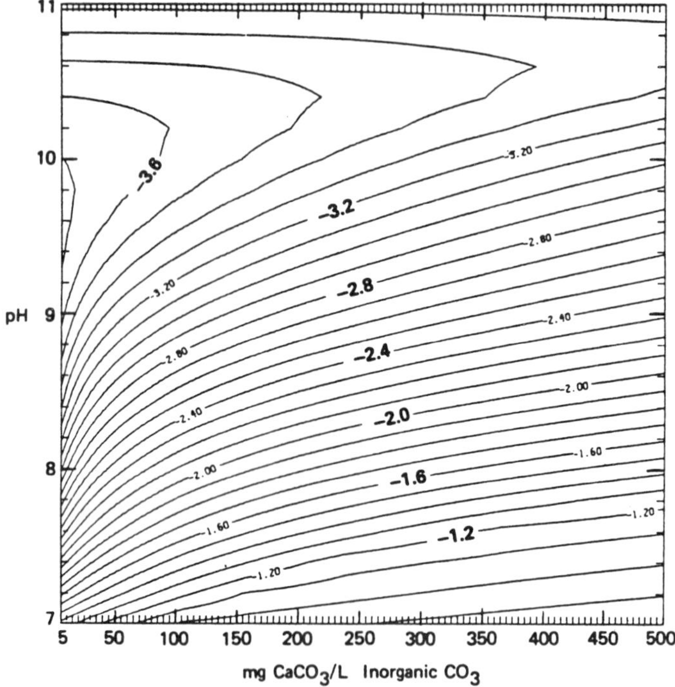

Figure 5. Contour diagram of copper (II) solubility in the
system Copper (II)-Water-Carbonate at 25°C and I = 0.005
mol/L. All points lie below the SMCL of 1 mg/L (0.00 in
diagram units). The contour interval is 0.1 log units.

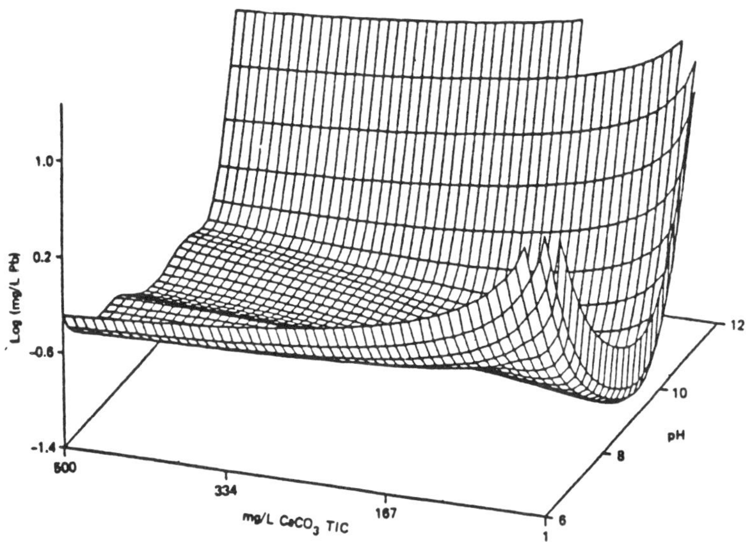

Figure 6.   Three-dimensional solubility diagram for lead (II)
corresponding to the conditions of Figure 1.

galvanized surface layer is of concern, because even at modest zinc solubilities the layer could be perforated if not "sealed" in time by the calcium carbonate coating.

In contrast to zinc and copper, lead (and soldered joints in copper pipe containing lead) could cause a serious problem (Figure 1). Assuming a pH range of 7.5 to 8.5 and a TIC concentration of approximately 200 to 400 mg $CaCO_3$/L (carbonate alkalinity of approximately 100 to 200 mg $CaCO_3$/L), the dissolved lead concentration could easily attain 0.16 to 0.25 mg/L upon overnight standing. Flowing samples could readily approach the MCL of 0.05 mg/L, given a sufficient length of pipe. Of additional concern is the fact that fresh leaded solder would generally be exposed immediately after house or building construction, or copper pipe repair. Leaded solder has also been noted for connections of plated brass household faucet assemblies (19). In some communities, lead is still either allowed or specified (20) for new service lines of small diameters (approximately 0.75 to 2 inches).

There are three special practical considerations in implementing a control program based on calcium carbonate saturation that should be noted. While the Langelier Index has been included as a parameter to be monitored in the EPA drinking water regulations (1), numerous researchers have pointed out that two waters of the same Langelier Index would not necessarily have the same potential mass of calcium carbonate to deposit (13, 18, 21). This value, the "calcium carbonate precipitation potential" (CCPP), is really the most enlightening index, and deserves to replace the Langelier Index as a control parameter.

The second consideration is that the identity of the calcium carbonate phase that precipitates is frequently the more soluble aragonite form, rather than calcite, the latter being the assumed phase in almost all calculations. Exactly what pipe, temperature, and water quality conditions govern this phenomenon have not been investigated to any great degree. To further complicate matters, a third form, vaterite, has been observed in some hot water systems (22). Log solubility constants vary from −7.91 for vaterite to −8.48 for calcite at 25°C and an ionic strength of zero. The solubility of each different form of $CaCO_3$ also varies in its response to temperature changes.

The third consideration is the fact that the most widely used values for the solubility constant of calcite at different temperatures have been shown to be in considerable error by studies over approximately the last decade (10, 23). Revised values based on the latest research have been presented (10, 23), and the widely used method of "total dissolved solids" correction originally developed in 1942 (24) has been revised (25).

pH ADJUSTMENT

Figures 1, 3, and 5 show that for a simple system consisting of the metal, water, and carbonate, the solubilities of lead (II), copper (II),

and zinc can all be greatly reduced by increasing pH into the range of 9 to
10. In this pH range, in cold or room-temperature waters, the solubilities
are controlled by hydrocerussite ($Pb_3(CO_3)_2(OH)_2$), tenorite (CuO),
and probably hydrozincite ($Zn_5(CO_3)_2(OH)_6$). Lead is particularly
sensitive to both pH and TIC, while pH is the dominant factor for copper
(II) and zinc at TIC levels above approximately 20 mg $CaCO_3$/L.

While hydrocerussite and tenorite form tight and adherent scale
layers, the films formed on zinc may not be as stable. Bachle et al.
suggest that the corrosion products in the cover layer have no
transport-inhibiting effect (26). Several studies have indicated
conversions of an initially precipitated zinc oxide or hydroxide to a basic
carbonate, and have indicated dehydration and phase changes in hot water
systems (27–29). Changes in the morphology of the corrosion product of
zinc with pH and TIC warrant continued investigation, in order to attempt
to prevent removal of the thin galvanized layer, even while dissolved zinc
remains below the SMCL level of 5 mg Zn/L. Essentially, no data is
available for the dependence of complex formation and solubility constants
of important lead (II), zinc, or copper (II) solids on temperature.

pH AND CARBONATE ADJUSTMENT

This treatment approach has been extremely successful in the case of
lead, both in laboratory (8, 9) and field studies (30). Many waters
contain sufficient carbonate to help limit lead (II) solubility, if only
their pH is increased. A low alkalinity does not necessarily mean a low
carbonate content, particularly when the pH is below 7. Because lead (II)
forms strong carbonate complexes, only waters of very low carbonate
concentrations are amenable to this treatment. Waters with high
alkalinities and neutral to basic pH values may have to have carbonate
removed to enable the attainment of the MCL.

Copper (II) also forms strong carbonate complexes, and its solubility
is increased by carbonate supplementation in the pH range of pH 7 to 10.5.
Copper (II) solubility remains below about 0.01 mg/L above pH 8.5, so
maintenance of the SMCL should not often be a problem. Of considerable
concern, however, is the possibility of rapid pitting (generally called
"Type I" pitting) and perforation of the pipe wall. No consensus has been
reached on the causes and mechanisms of pit propagation. Mattsson and
Fredericksson (6) examined numerous corrosion products from systems with
pitting, and concluded that in waters high in sulfate and chloride, basic
copper (II) chloride and sulfate salts formed, rather than just tenorite
and malachite. They suggested the ratio of bicarbonate to sulfate, or
chloride, or both, could be important, but the effect might be
pH-dependent. In this case, carbonate supplementation in addition to pH
adjustment should prove to be beneficial.

Except for the pH range of 9 to 10.8 at high TIC concentrations, and
the extremely low TIC region (10 mg $CaCO_3$/L) in the pH range of 8 to
10.5, zinc solubility is relatively insensitive to TIC. Thus, the benefit

of carbonate supplementation might be analogous to the concept of calcium carbonate saturation control, wherein an additional mass of carbonate might tend to form a more effective film. There is no published experimental verification of this, however, and an improvement in corrosion resistance with continually increasing TIC should not be presumed. Here, there may be a difference in behavior as the underlying steel alloy layer becomes exposed, and the corrosion reactions of iron take over.

ORTHOPHOSPHATE ADDITION

In 1970, Murray (31) developed a corrosion inhibitor chemical from a mixture of zinc sulfate, monosodium orthophosphate, and sulfamic acid (31). He stated that the mechanism of corrosion inhibition was the formation of a protective coating of $Zn_3(PO_4)_2(s)$ because ". . . zinc phosphate is insoluble in water at all concentrations. . . ." That is obviously false, because zinc orthophosphate solid does have a measurable aqueous solubility constant (Table 3), but the chemical combination has become a useful corrosion inhibitor.

Lead (II) forms at least one orthophosphate solid of low solubility in realistic drinking water conditions, which can serve as the basis for corrosion control (9, 17, 30, 32, 33). Figure 2 shows that the predicted equilibrium lead (II) solubility can now fall below the MCL at low carbonate concentrations, with a dosage of only 0.5 mg PO_4/L. Experiments with both lead and galvanized pipes at U.S. EPA have demonstrated that supersaturation with zinc orthophosphate is not necessary for lead solubility control because the lead forms a passivating lead orthophosphate film (30). Investigations by the Water Research Centre in England (32) have indicated the solid to be hydroxypyromorphite, $Pb_3(PO_4)_2OH$. Successful field applications in England have been reported by Sheiham and Jackson (32).

The passivating action of the orthophosphate depends on the pH, the TIC concentration, the orthophosphate concentration, and temperature. No data is available on the effect of the latter. On new lead surfaces, a very thin, smooth film is formed. Taylor (34) has suggested that the orthophosphate might not be as effective if the pipe surface is already covered with corrosion products and deposits (34).

If the solubility constant for $Cu_3(PO_4)_2 \cdot 2H_2O$ is accurate, and if it is a realistic phase that would precipitate in drinking water systems, then orthophosphate addition would be of little benefit for copper level control. Solubility calculations were performed for orthophosphate levels up to 2 mg PO_4/L, and no precipitation of a copper (II) orthophosphate solid was predicted. If either or both of the solubility constants used for tenorite and malachite are too low, then this conclusion would have to be reevaluated. A zinc orthophosphate coating on the copper could only be deposited if its solubility product would be exceeded. High concentrations of zinc and orthophosphate might be required, depending upon the pH.

Figure 4 shows the solubility diagram for zinc when a dosage of 0.5 mg PO_4/L is added, and $Zn_3(PO_4)_2 \cdot 4H_2O$ (-hopeite) is the probable solid orthophosphate precipitate. At pH values under 7.5 and at low TIC concentrations, some solubility suppression is likely. Increasing the orthophosphate concentration would extend the stability domain of -hopeite. Zinc leaching from galvanized pipe might be slowed or limited by the mass-action principle, if zinc ion were added to the water along with the orthophosphate inhibitor. More research in this area is definitely needed.

Four advantages to orthophosphate addition plus pH adjustment are apparent. First, for at least lead and zinc, the same solubility can be attained at a lower pH than using pH plus carbonate alone. This assists disinfection and slows organic halogen formation in chlorinated supplies. Second, orthophosphate may be beneficial to the iron-zinc alloy substrate beneath the galvanized layer, yielding superior protection where the surface layer is finally penetrated. This area needs significant investigation. Third, the same treatment strategy might also benefit asbestos cement, cast iron, or cement mortar-lined distribution pipe lines. A fourth advantage is that for waters with high carbonate contents, a significant reduction in lead solubility would be possible without softening or decarbonation.

POLYPHOSPHATE ADDITION

Polyphosphates are strong chelating agents for calcium, magnesium, iron, and other trace metals. While they have useful properties for reducing scaling, cleaning tuberculation from distribution mains, and reducing "red water" complaints through sequestration of Fe^{2+} ion, they have not been proven useful in reducing the solubility of zinc, copper, or lead. Even though they may be somewhat effective relative to nontreatment, particularly in acidic waters, alternative treatment strategies will virtually always prove superior in reducing metal levels. Any joint use of polyphosphate chemicals with another treatment (such as to prevent scaling at high pH, or with sodium silicate) must be accompanied by thorough monitoring of trace metal concentrations. This is necessary because the actual polyphosphate ligands have not been identified, and their complex formation constants have not been determined, making even semiquantitative estimates of effect through solution modeling calculations very unreliable.

SILICATE ADDITION

Silicate addition has been demonstrated to be an effective procedure for corrosion inhibition of galvanized steel pipes, particularly at high temperatures. Studies by Glasser and Lachowski of the structures of silicate solutions (35) concluded that there is essentially no difference between those prepared by the dissolution of anhydrous sodium metasilicate or by salts with sodium orthosilicate groups. Therefore, either type of additive should provide similar results for corrosion inhibition.

Early studies with galvanized and brass drinking water pipes indicated that the silicate additive (8 to 12 mg SiO_2/L) helped form amorphous films binding corrosion products of the metal, and additionally enmeshing iron oxide and organic particles present in the water supply (36). Thus, an initial corrosion product needed to be present in order for the silicate to work effectively. Similar conclusions may be drawn about the silicate action on limiting the dissolution of zinc electrodes, zinc oxide, and zinc dust through the formation of adsorption compounds that control zinc ion diffusion (37). Silicate treatment may, therefore, bridge the gap between control of zinc levels in solution by the precipitation of basic zinc carbonate, zinc hydroxide, or zinc oxide, and sustaining the life of the zinc coating by the formation of a uniform and adherent passivating film.

An indication of this potential has been given by studies done at the Illinois State Water Survey with simulated domestic hot water systems (38, 39). In waters of low carbonate content, a film of zinc pyrosilicate was apparently formed at a pH of 8 and a dosage of 8 mg SiO_2/L at 60°C (140°F). However, in waters with higher carbonate contents, an adherent scale of basic zinc carbonate mixed with silicate was found (38). Further experiments were conducted to better define the optimum pH and silicate dosage in a water of approximately 100 mg $CaCO_3$/L total hardness, 35 to 55 mg $CaCO_3$/L total alkalinity, and 8 mg SiO_2/L natural silicate (39). An initial high dosage of silicate (28 mg SiO_2/L total) was found to be useful to provide more immediate corrosion inhibition and less bulky scale formation, but the optimum dosage for this water quality was about 18 mg SiO_2/L at a pH of approximately 8.5. Corrosion inhibition in these experiments was monitored by the use of standardized ASTM pipe coupon inserts for weight-loss determinations (40). Silicate requirements tended to be less as the pH was increased by caustic soda additions. The authors concluded that in the hot water systems, the optimum silica concentration and pH may be different for waters of higher alkalinity, where more basic zinc carbonate corrosion product would be formed, whereas primarily hemimorphite ($Zn_4Si_2O_7(OH)_2 \cdot H_2O$) was found in this study. Little or no calcium carbonate deposition was detected. They also indicated that other water quality factors, such as chloride, sulfate, calcium, and magnesium, could change the dosage requirements and could influence properties of the scale.

In a study of two fairly soft waters in Seattle, sodium silicate addition at a pH of 8.0 considerably reduced the corrosion rates of black steel pipes compared with other treatments investigated (41). After two or three months of exposure of galvanized steel pipe coupons, the silicate treatment showed slightly inferior results to that attained by lime plus soda ash. Obviously, the interplay of pH, silicate dosage, and initial corrosion product formed on the pipe is of critical importance at low temperatures as well as in hot water systems. Further investigation is necessary before treatment guidelines can be given with much confidence.

Silicate addition has not been found to be particularly effective for copper pipe, although few studies have actually been conducted. No apparent reaction was found (38) between copper metal and a soft water containing 8 mg SiO_2/L at pH 8.4 at 60°C (140°F). The "control" experiment in this study was at pH 6.7, making a direct comparison of the effect of pH adjustment alone to the different inhibitors tested very difficult. Also, in waters tested having a high hardness, calcium carbonate precipitation was suspected, making resolution of the role of the sodium silicate inhibitor difficult. The scales seemed to offer good corrosion resistance, however.

In the Seattle study (41), silicate addition proved to be advantageous only in keeping even the initial corrosion rates of the copper specimens low. After three months of exposure, alternate treatments such as lime plus soda ash or lime plus zinc orthophosphate were equivalent or superior.

Some experimental tests have been conducted with lead pipe (30, 32), and observations of the inhibitory effect of naturally occurring silica have been summarized (30). After approximately 8 to 9 months of exposure to 20 mg SiO_2/L, lead levels were slowly reduced in a soft, low-alkalinity water at pH 8.2. In systems with both new galvanized and lead pipes and 10 mg SiO_2/L concentration, lower lead levels were observed with continued exposure. These levels were at approximately the 0.05-mg Pb/L MCL, but presumably the pH and TIC content of the water would be important variables to consider before extrapolating results. These experiments tend to support the idea of the slow formation of a film on new pipe, but which might become effective over the long term as a "cementing" agent for other corrosion products. Corrosion films on lead pipe always tend to be very thin and uniform, unlike those on galvanized pipe. Sheiham and Jackson (32) conducted some experiments with low alkalinity moorland waters and concluded that 10 mg SiO_2/L had little effect on lead dissolution.

CONCLUSIONS

Case studies and solubility calculations for zinc, copper (II), and lead (II) in drinking water show that several control strategies have been proven to be adequate for the attainment of current MCLs and SMCLs. The most difficult metal to control is lead, and for some water chemistries flushing of the lines before taking water for human consumption may be required if optimization of treatment for lead corrosion control cannot be accomplished. Lowering of the lead MCL to 0.025 or 0.01 mg/L would cause considerable difficulty in light of current knowledge about available inhibitors.

In order to develop efficient and accurate treatment techniques, considerable effort must be expended in the future to understand the mode of action of phosphate and silicate inhibitors, as well as the chemistry of lead, copper, and zinc in response to changes of pH, TIC, and temperature. This research is vital both to ensure the toxicological purity of the water, and to extend the life of plumbing materials of diverse composition in the same water distribution system.

REFERENCES

1. EPA. 1976. National Interim Primary Drinking Water Regulations. EPA-570/9-76-003. U.S. Environmental Protection Agency Office of Water Supply.

2. EPA. 1980. Interim Primary Drinking Water Regulations; Amendments. Federal Register 45(168):57332 (August 27, 1980).

3. EPA. 1979. National Secondary Drinking Water Regulations. EPA EPA-570/9-76000. U.S. Environmental Protection Agency Office of Drinking Water.

4. Ryder, R.A. In preparation. Corrosion Inhibitors. In: Internal Corrosion of Drinking Water Systems. AWWARF.

5. Kuch, A., and I. Wagner. 1983. A mass transfer model to describe lead concentrations in drinking water. Water Res. 17(10):1301.

6. Mattsson, E., and Fredericksson. 1968. Pitting corrosion in copper tubes — cause of corrosion and counter-measures. Br. Corros. J. 3(9):246 (Sept. 1968).

7. Schock, M.R. 1980. Computer modeling of solid solubilities as a guide to treatment techniques. Seminar on Corrosion Control in Water Distribution Systems. U.S. EPA, Cincinnati, May 20-22.

8. Schock, M.R. 1980. Response of lead solubility to dissolved carbonate in drinking water. Jour. AWWA 72(12):695. Errata 73(3):36 News (1981).

9. Schock, M.R., and M.C. Gardels. 1983. Plumbosolvency reduction by high pH and low carbonate; solubility relationships. Jour. AWWA 75(2):87.

10. Schock, M.R. and C.H. Neff. 1982. Chemical aspects of internal corrosion; theory, prediction, and monitoring. Proc. AWWA Water Quality Technology Conference X, Nashville, Tennessee, Dec. 5-8.

11. Schock, M.R., and R.W. Buelow. 1981. The behavior of asbestos-cement pipe under various water quality conditions: Part 2, theoretical considerations. Jour. AWWA 73(11):609.

12. Stumm, W. 1956. Calcium carbonate deposition at iron surfaces. Jour. AWWA 48:300 (1956).

13. Stumm, W. 1960. Investigations of corrosive behavior of waters. Trans. ASCE - San. Engrg. Div. 86:27.

14. Sontheimer, H., W. Kolle, and V.L. Snoeyink. 1981. The siderite
     model for the formation of corrosion-resistant scales. Jour. AWWA
     73(11):572.

15. Kuch, A., and H. Sontheimer. 1984. Corrosion in Iron and Steel
     Pipes, or the "Protective-Layer-Story." DVGW Forschungsstelle am
     Engler-Bunte-Institut, Postfach 6380, 7500 Karlsruhe 1, West Germany.
     Unpublished.

16. Singley, J.E. 1981. The search for a corrosion index. Jour. AWWA
     73(11):579.

17. Gardels, M.C., and M.R. Schock. 1981. Corrosion indices: invalid or
     invaluable? Proc. AWWA Water Quality Technology Conference IX,
     Seattle, Washington. EPA 600/D-82-108.

18. Rossum, J.R., and D.T. Merrill. 1983. An evaluation of the calcium
     carbonate saturation indexes. Jour. AWWA 75(2):95.

19. Samuels, E.R., and J.C. Meranger. 1984. Preliminary studies on the
     leaching of some trace metals from kitchen faucets. Water Res.
     18(1):75.

20. Chapters of the Municipal Code of Chicago Relating to Plumbing, with
     Amendments to October 19, 1978. Index Publishing Corp.

21. Loewenthal, R.E., and G.v.R. Marais. 1976. Carbonate Chemistry of
     Aquatic Systems: Theory and Applications. Ann Arbor Science.

22. Carlson, W.D. 1983. The polymorphs of $CaCO_3$ and the
     aragonite-calcite transformation. In: R.J. Reeder (ed.),
     Carbonates: Mineralogy and Chemistry, Reviews in Mineralogy, Vol. 11,
     Mineralogical Society of America.

23. Plummer, L.N., and E. Busenberg. 1982. The solubilities of calcite,
     aragonite and vaterite in CO_2-H_2O solutions between 0 and 90°C,
     and an evaluation of the aqueous model for the system
     $CaCO_2$-CO_2-H_2O. Geochim. Cosmochim. Acta 46:1011.

24. Larson, T.E., and A.M. Buswell. 1942. Calcium carbonate saturation
     index and alkalinity interpretations. Jour. AWWA 34:1667.

25. Schock, M.R. 1983 (in press). Temperature and Ionic Strength
     Corrections to the Langelier Index (Revisited). Submitted manuscript.

26. Bachle, A., et al. 1981. The corrosion of galvanized and unalloyed
     steel pipes in drinking water of different hardness and neutral salt
     content. (In German.) Werkstoffe und Korrosion 32:435.

27. Anderson, E.A., and M.L. Fuller. 1939. Corrosion of zinc. Metals
     and Alloys (Sept. 1939).

28. Gilbert, P.T. 1952. The nature of zinc corrosion products. *Jour. Electrochem. Soc.* 99(1):16.

29. ILZRO. 1967. *Surface Characteristics of Zinc in Aqueous Media.* Final Report. ILZRO Project ZE-20 (May 1967). International Lead Zinc Research Organization, Inc.

30. Schock, M.R. 1984 (in preparation). The corrosion and solubility of lead in drinking water. In: *Internal Corrosion of Drinking Water Systems*, AWWARF.

31. Murray, W.B. 1970. A corrosion inhibition process for domestic waters. *Jour. AWWA* 62(10):659.

32. Sheiham, I., and P.J. Jackson. 1981. The scientific basis for control of lead in drinking water by water treatment. *Jour. Inst. Wtr. Treatment Engr. and Scientists* 35(6):491.

33. Gregory, R., and P.J. Jackson. 1983. Reducing Lead in Drinking Water. *Water Research Centre Report 219-S.* Stevenage Laboratory, Elder Way, Stevenage, Herts SG1 1TH, England (May-June 1983).

34. Taylor, F.B. 1980. Metals — Corrosion or Dissolution? *Proc. of the Seminar on Corrosion Control in Drinking Water Systems*, NEWWA (Mar. 24-25, 1980).

35. Glasser, L.S.D., and E.E. Lachowski. 1980. Silicate Species in Solution. Part 1, Experimental Observations. *J.C.S. Dalton Trans.:* 393-398.

36. Lehrman, L., and H.L. Shuldener. 1951. The role of sodium silicate in inhibiting corrosion by film formation on water piping. *Jour. AWWA* 43(3):175.

37. Glasser, L.S.D., et al. 1978. The reaction of zinc oxide and zinc dust with sodium silicate solution. *J. Appl. Chem. Biotechnol.* 28:799.

38. Lane, R.W., et al. 1973. Silicate treatment inhibits corrosion of galvanized steel and copper alloys. *Mat. Perform.* 12(4):32.

39. Lane, R.W., et al. 1977. The effect of pH on the silicate treatment of hot water in galvanized piping. *Jour. AWWA* 69(8):457.

40. ASTM. 1976. Standard Methods of Test for Corrosivity of Water in the Absence of Heat Transfer (Weight Loss Methods). D2688-70, *ASTM Standards, Part 31*:141.

41. Ryder, R.A. 1978. Methods of Evaluating Corrosion. *Proc. AWWA Water Quality Technology Conference VI*, Louisville, Kentucky, Dec. 3-6 (1978).

DRINKING WATER REGULATIONS AND THEIR IMPACT ON MATERIALS USED IN WATER SYSTEMS

James R. Boydston

Manager, Drinking Water Program
Oregon State Health Division
P.O. Box 231
Portland, Oregon 97207

Regulations and plumbing code requirements have a very definite influence on the materials used in water systems and house plumbing. Such regulations are necessary to provide for protection of the public health where the combination of water quality and piping materials may result in hazardous conditions. One of the problems of greatest current concern is the presence of heavy metals in drinking water resulting from lead or other piping materials being used to convey waters that are highly corrosive.

A number of regulations have been developed at both federal and state levels in an attempt to prevent health hazards. The U.S. Environmental Protection Agency (EPA) is currently in the process of adopting national revised primary drinking water regulations (1). The current interim regulations specify maximum contaminant levels for microbiological, inorganic, and organic chemicals and radionuclides (2). The maximum contaminant level (MCL) means "the maximum permissible level of a contaminant in water which is delivered to the free flowing outlet of the ultimate user of a public water system." These MCLs are currently being reviewed and in a number of instances may be significantly lowered.

In addition to the MCLs, the interim regulations also include requirements to identify the presence of specific materials in distribution systems and to monitor for corrosivity. This requirement stems from increasing concern about corrosion byproducts and particularly the much higher levels of metals observed in corrosive water when that water stands for some time in house plumbing systems. The interim regulations require that all community water systems sample their water twice during a one-year period and report the Langelier index as an indication of the corrosivity of the water. This testing is required in order to enable the primary enforcement agency to determine which water supply systems should initiate corrosion control measures. The approach being considered for the revised primary drinking water regulations is to set specific monitoring requirements for corrosion byproducts such as lead and cadmium. The systems known to have corrosive water will be required to take sufficient samples to insure that MCLs for the corrosion byproducts will not be exceeded. Under this interim regulation the states were encouraged to implement corrosion control measures at systems distributing corrosive waters. Much has been written about the shortcomings of the Langelier index as a measure of corrosivity but, since no better test has been developed, this remains the index of choice.

EPA has also developed Water Supply Guidance Number 73 on sampling techniques to be used for compliance monitoring for metals in drinking water (3). These sampling techniques are not federally enforceable but are intended to provide a standardized method for testing for corrosion byproducts. This guidance provides for collection of three samples: from a house faucet to represent the water that has been standing overnight in the house plumbing, from the service line to the house, and from the distribution system. Rather than using this sampling protocol to determine MCL violations, it is to be used more as an indication of the potential for exposure to higher levels of corrosion byproducts than would be identified from the usual primary inorganics testing.

In addition to federal and state drinking water regulations, plumbing code requirements also influence materials used in house plumbing systems. The plumbing codes of most states are based upon some national model such as the Uniform Plumbing Code. The Oregon Code with which I am familiar is based upon the Uniform Plumbing Code of the International Association of Plumbing and Mechanical Officials with local variations to meet Oregon requirements (4). One interesting feature of the Oregon Plumbing Code is a provision that the Administrator of the Health Division may prohibit the use of any pipe, solder, fillers, or brazing materials which do not conform to product acceptability criteria adopted by the Health Division. The Oregon Code specifies that the approved materials used in house plumbing may include brass, copper, cast iron, galvanized malleable iron, galvanized wrought iron, galvanized steel, polybutylene (PB), chlorinated polyvinyl chloride (CPVC), or other approved material. Polyethylene (PE) and polyvinyl chloride (PVC) water pipe may be used only for cold water systems outside of buildings. Lead pipe is not permitted in potable water systems.

Oregon law was changed in 1979 to give the Health Division more direct influence over plumbing regulations. This was due in part to problems in the 1970s with galvanized pipe imported from overseas that was not manufactured to American standards and resulted in serious taste and odor problems, rapid corrosion, and some potential for gastrointestinal upset. The State Legislature amended the law relating to regulation of plumbing to require that all potable water pipe sold in Oregon be clearly marked with the identification of the manufacturer and plant of origin. This permits easier tracking of inferior pipe.

Recent regulatory changes throughout the country have taken a number of directions. This may be due in part to a lack of federal standards for corrosion testing and MCLs for corrosion byproducts. A number of cities have become concerned about the potential health effects of corrosion and corrosion byproducts and have funded engineering and medical studies to

evaluate the impact of corrosion. The problem with lead in drinking water
is certainly not new in this country. In an 1845 report to the City of
Boston (5), it was concluded that: "Considering the deadly nature of lead
poison, and the fact that so many natural waters dissolve this metal, it is
certainly the cause of safety to avoid, as far as possible, the use of lead
pipe for carrying water which is to be used for drinking." The advice was
not heeded, however. In 1920 another Boston report (6) stated: "If water
supplying such works be found to attack lead, the best plan is to replace
them gradually, meanwhile protecting consumers by proper treatment of the
water supplied through existing services." Boston's approach to resolving
the problem was to provide treatment to reduce the corrosivity of water.
Tests run on water from consumers' taps in Boston before and after
implementation of treatment showed a significant drop in lead concentration
from above 0.05 mg/L to consistently below that level. This is
accomplished by pH adjustment to approximately 8.5 using sodium hydroxide.
Much lead pipe plumbing remains in use in the Boston system.

In other parts of the country, notably Seattle and Portland, some lead
service connections remain, but a significant source of lead appears to be
from lead-based solders used to join copper pipe and fittings.

The State of Delaware is the only state to date that has elected to
ban the use of lead-based solder for potable water systems. The Delaware
plumbing code requires that soldered joints in copper tubing and fittings
shall be made with solder conforming to ASTM B32, provided that the lead
content shall not exceed 0.20 percent for potable water systems. This in
effect will require that tin-antimony or tin-silver solders be substituted
for the usual 50/50 tin-lead solder.

The City of Seattle has used two soft surface-water sources for many
years. Complaints about rusty water or blue stains and resulting taste
problems led the City of Seattle to sponsor an extensive engineering study
of corrosion and corrosion effects in the Seattle water system (7). The
study confirmed that both sources of water were corrosive and exposed the
Seattle customers to corrosion byproducts such as lead in addition to the
esthetic effects. Seattle's approach to the problem was two pronged — the
adoption of a number of amendments to the Uniform Plumbing Code and the
provision for treatment of water to neutralize corrosivity. Amendments to
the Plumbing Code included the banning of lead solder, the requirement for
an approved dielectric fitting when connecting piping of dissimilar
materials, and the approval to use PB or CPVC as hot or cold water piping
in buildings not exceeding four stories in height. In addition the City
now prohibits the use of house plumbing systems for electrical ground.

The treatment applied in Seattle consists of the addition of lime and
sodium bicarbonate to increase the pH and alkalinity of both sources.
Incidentally, Seattle's engineers estimated that there would be a net
annual savings to the consumer, looking at a benefit/cost ratio of 5.4:1
when comparing the total annual cost of the treatment versus the projected
net annual consumer plumbing maintenance cost savings.

Portland also had an engineering study performed which confirmed that the untreated water from Bull Run River is also highly aggressive (8). Although the Portland system still contains roughly 10 percent lead services or pigtails, the engineers felt the major health risk was from lead solder in newer copper-plumbed systems. A supplemental study was done on some non-copper-plumbed houses that were known to have lead pigtails, and it was concluded that the lead spike that occurred in water standing within the pigtails would be dispersed below MCL levels before reaching the kitchen tap. It was concluded then that Portland should continue to replace lead pigtails as they were found and that banning lead solder in new copper plumbing would be sufficient to protect the public health.

Portland's first approach to the problem was to consider banning lead solder by ordinance in the City of Portland, but later requested that the State Health Division ban lead solder statewide since Portland serves water to a number of water districts outside the city limits. A city ordinance would have no effect within these districts. The Health Division was concerned that, although banning lead solder would solve a major part of the problem with new construction, there would still be concern about the health risk exposure from existing plumbing systems. The state held a number of informational hearings to receive comments on a proposed lead solder ban and other options. Much comment from the plumbing industry centered on the difficulty of using the alternative solders due to their narrower working temperature range and the consequences of greater expense for new construction. Many comments reflected a concern that such a proposal would be unenforceable and was not necessary in many areas of the state where hard groundwater sources are used. Some testimony was received from communities that already provided full treatment, including pH correction, that the lead solder ban was unnecessary.

The State of Oregon is still considering a number of alternatives, including a statewide ban on lead solder with certain regions of the state eligible for an exemption from the ban, and a monitoring program to improve identification of communities that have corrosive water followed by a requirement that those communities prepare plans for corrosion control. Oregon will shortly draft a proposed rule incorporating the alternatives and will schedule a series of formal hearings to receive testimony prior to adoption or modification of the rule.

It is obvious that regulations and plumbing code requirements take a number of directions as concerned officials search for methods to safeguard against potential health risks. There are two major difficulties with the problem of corrosion byproducts. First, in the case of lead and perhaps other heavy metals, the effects are accumulative and chronic exposure to low levels may result in health impacts that are difficult or impossible to measure. Studies done by Dr. Needleman of the Harvard Medical School (9) confirm that children exposed to low lead dose levels performed less well than children not so exposed, even though neither group showed any other symptoms of lead poisoning. A second difficulty with the problem of lead in water is the lack of adequate methods for measuring corrosivity.

Under these circumstances regulatory officials have an obligation to take a conservative approach and require all reasonable methods to minimize or to prevent any uptake of lead through drinking water. Since there are reasonable treatment or control methods available, communities should voluntarily be providing corrosion control. Such treatment also will have esthetic benefits that will more than offset treatment or control costs.

Although there has been a great amount of research done on corrosion, the water supply industry still has no simple reliable test to measure it. Coupon testing has been suggested as an accurate measurement of corrosion, but time required for the testing makes this impractical. I would challenge the research community to develop a standard method for measuring corrosivity by exposing a volume of water to a specimen of lead, leaving the water standing for a selected time, then measuring the lead uptake in the water. It appears a reproducible laboratory test could be developed that would approximate water standing in a plumbing system. This type of test is urgently needed to help provide answers to questions being raised by a concerned public.

REFERENCES

1. Federal Register Vol. 48, No. 194, October 5, 1983. National Revised Primary Drinking Water Regulations, Advance Notice of Proposed Rulemaking.

2. U.S. EPA. 1976. National Interim Primary Drinking Water Regulations. EPA-570/9-76-003. U.S. Environmental Protection Agency.

3. U.S. EPA. 1982. Guidance for Monitoring and Sampling Techniques to Determine Corrosion Products, Including Lead in Water Supply Distribution System. Water Supply Guidance Number 73. U.S. Environmental Protection Agency.

4. International Association of Plumbing and Mechanical Officials. 1979. Uniform Plumbing Code.

5. Report of Commissioners Appointed by Authority of the City Council to Examine the Sources From Which a Supply of Pure Water May Be Obtained for the City of Boston. J.H. Eastburn, City Printer. Boston, Massachusetts, 1845.

6. R.S. Weston. 1920. Lead poisoning by water and its prevention. JNEWWA 34:239.

7. Kennedy Engineers. 1978. Internal Corrosion Study, Phases 1, 2, 3, Seattle Water Department.

8. James M. Montgomery, Consulting Engineers. 1982. Internal Corrosion Mitigation Study, Final Report, Portland Bureau of Water Works.

9. Needleman, H.L., C. Gunnoe, A. Leviton, R. Reed,
    H. Peresie, C. Maher, and P. Barrett. 1979. Deficits in psychologic
    and classroom performance of children with elevated dentine lead
    levels. New England Journal of Medicine 300 (13):689.

EUROPEAN DEVELOPMENTS IN USE OF
PLUMBING MATERIALS

Ivo Wagner

Head, Corrosion and Material Testing Section
Engler-Bunte Institute
University of Karlsruhe
Karlsruhe, Germany

INTRODUCTION

The plumbing materials used in Europe during the last century were
lead, copper, and galvanized steel. Recently plastic materials, especially
cross-linked polyethylene, came on the market for residential
drinking-water installations. Depending on the distribution method,
individual countries made greater or lesser use of different plumbing
developments.

Changes in the quality and use of plumbing materials were due to
changes in price or technical quality (mainly due to corrosion), and health
considerations. With these changes, use of lead plumbing in Europe
decreased nearly to nothing, and the installation of cross-linked
polyethylene pipes in a double-tube system with metallic joints became
commonplace.

Although the cost of plumbing systems in new-house installations,
mainly in Germany, increased over the last 30 years (with the ratio of
galvanized steel to copper pipes changing from 6:1 to 3:2), the use of lead
was ruled out because of health problems. Polyethylene pipe was developed
mainly to correct local corrosion problems caused by metallic pipe. More
detailed information on the different materials, their quality standards,
and their characteristics are given below.

LEAD

Lead plumbing and lead distribution systems were used in Europe to a
great extent when central water-distribution systems were being developed,
in the later decades of the 19th century. In spite of warnings to be found
in the literature since the end of the 18th century, lead was the preferred
plumbing material up until the late 1930s.

The regulations covering lead varied by country. For example, to reduce lead uptake by water, Prussia recommended that lead had to be free from other metals. The Netherlands passed a regulation requiring a layer of tin inside lead pipes and, when that was discovered to be hazardous, amended the regulation to require the tin content to be 99.5 percent.

In other countries, mainly in some German states, lead was banned completely, beginning with Wuerttemberg in 1878 (1). Lead was finally ruled out in Germany in 1972 with the introduction of the new German Standard DIN 2000. (A general survey of this development is given in Table 1.)

Similar developments took place throughout Europe at a pace dependent on the national regulations and possibilities. In the United Kingdom, for example, a recommendation against the use of new lead pipes was given in the early 1970s. The last new lead plumbing systems were installed in 1976. Table 2 shows the development of recommendations in Great Britain.

These very restrictive measures concerning lead plumbing were supported, mainly with respect to the old distribution systems, by a lot of research into water quality and levels of lead in drinking water. Numerous studies and surveys have been performed and published in the last 10 years (3,4,5). The main conclusion reached was that the level of lead in both soft and hard waters nearly everywhere exceeded the Maximum Acceptable Concentration (MAC) values, often to a high extent. The solution to this problem was either replacement of lead by other materials or treatment of water to reduce lead solubility (6).

The questions to be answered in this case were addressed by studies performed in the United States (7) and the United Kingdom (8) on the stability or solubility of lead compounds. Additional data were gathered (9) by monitoring and calculating lead levels as a function of standing time; the maximum solubility of lead in water was reached after long standing time.

With respect to the arguments of European toxicologists (10, 11) that even low lead concentrations have to be seen as a cumulative poison, the longtime uptake should be evaluated. With this goal, further research was performed on consumption patterns of households in the United Kingdom and Germany (12, 13). Lead levels were compared depending on maximum concentration, kind of pipe, pipe length and diameter, flow velocity, and standing time. These data allow for the calculation of general and individual lead intake by a minimum of monitoring and analysis.

COPPER

The use of copper pipes was generally unrestricted due to health considerations. Nevertheless, blue or green discoloration of water due to stagnation (standing time) — mainly in acid waters — demonstrates that copper pipe can cause heavy metal uptake in water after a long stagnation

TABLE 1

LEAD: STANDARDS AND RECOMMENDATIONS
IN FEDERAL REPUBLIC OF GERMANY

YEAR(S)	REGULATION
1878-1972	Ban on lead pipes for distribution and house installation in different German states, beginning with Wuerttemberg (1878)
1972-present	Total ban for new installations in lead (DIN 2000)
1980/81-present	Study on Drinking Water and Lead (DVGW-Research Institute, Karlsruhe) Results: • Connecting pipes to be replaced depending on pipe length and water quality. • In house installations, MAC (40 ppb) is regularly exceeded. • Flushing and orthophosphate dosage may be used as immediate measure. • House installations in lead must be replaced as quickly as possible.

TABLE 2

LEAD: RECOMMENDATIONS AND MEASURES
IN GREAT BRITAIN

YEAR(S)	REGULATION
1976–present	Lead plumbing not further in use for new plumbing systems
1976–78	Countrywide survey on "Lead in Drinking Water" (DoE)
from 1980	Drinking water treatment: ● by pH increase in soft water ● by o-phosphate dosage in hard water
from 1982	Survey on consumption patterns in households (WRC) General exception in Great Britain: No pressurized systems in house installations; roof tanks, only kitchen tap under pressure

time. Water treatment by pH adjustment was therefore recommended in order
to reach drinking-water standards. As a result of corrosion problems, the
development of German Standards on copper pipe qualities was recommended.

Contrary to British and North European recommendations, the German
experience with copper pipe allowed for a certain level of carbon or
lubricant in the pipes. Since copper was used normally only in soft waters
or in hot-water installations, this allowance worked well. However, with
an increasing amount of copper-plumbed cold-water systems in regions with
hard groundwater, Type I pitting began to occur more frequently and finally
accounted for more than 70 percent of all corrosion cases (14). The
standards on lubricant, mainly carbon, on the inner pipe surfaces were
changed and the permissible levels decreased sharply. As a consequence of
that change, more hot (hard) soldering took place; after a while, pitting
in cold-water pipes increased once again. Thus the permissible carbon and
lubricant concentrations have been reduced again and a new surface
treatment on soft copper pipes has been introduced. These new pipes, which
have been in use for about 3 years, seem to meet the necessary
specifications; only a few cases of copper pipe failures have been reported.

The development of the German Standards governing copper pipe is given
in Table 3.

Although the use of copper pipe has presented no health problems, the
use of copper pipe in house installations is a concern due to the soft
soldering used to join the pipe. Most solders were tin/lead or
tin/antimony alloys or mixtures with differing lead content.

In the previously cited survey (2) on lead in the U.K., nearly 3,000
houses were monitored; with a certain astonishment, analysts registered
concentrations of lead in standing waters of copper-plumbed systems that
were higher than in lead-plumbed systems. Detailed research at the Water
Research Centre (15) in the U.K. and at KIWA (16) in the Netherlands has
shown that soft solders with lead content allow a high quantity of that
lead to migrate.

Similar research in Germany (17) studied the migration of lead out of
soft solders and of cadmium out of hard solders. The results were not
considered to be a problem of heavy metal uptake by water but, interpreting
the data correctly, it must be accepted that under actual pH conditions,
lead and cadmium uptake could become a severe health problem like that
found in U.K. and the Netherlands.

However, standards on solders have changed now in the U.K., the
Netherlands, and Germany. Soft solders now have to be free of lead; this
requirement is most important to the British plumbing systems, which were
normally installed with soft solder. The development of standards in
Europe is illustrated by the change in German Standards for soft and hard
solders. Table 4 shows the changes in the allowed solders, comparing the
standards for 1972 and 1983. Note that the change to solders that are free
of cadmium, antimony, and lead is complete.

TABLE 3

<u>DEVELOPMENT OF STANDARDS FOR COPPER PIPE
IN DRINKING WATER INSTALLATIONS</u>[1]

1972	1981	1983
Cu-hard	Cu-hard	Cu-hard
Lubricant (10 mg/dm^2)	Lubricant (4 mg/dm^2)	Lubricant 1 mg/dm^2 (diam. 2") 2 mg/dm^2 (diam. 2")
Cu-soft	Cu-soft	Cu-soft
Carbon 2 mg/dm^2	Carbon 0.8 mg/dm^2	Carbon 0.2 mg/dm^2 (no carbon films allowed)

[1]Derived from DVGW-Arbeitsblatt GW 392, Standard of German Gas and Water Association.

TABLE 4

DEVELOPMENT OF STANDARDS FOR SOLDERS ON COPPER PIPES
IN GAS AND DRINKING WATER INSTALLATIONS[1]

SOLDER	1972	1983
Hard solders	Ag 40 Cd Ag 30 Cd Ag 2 P	Ag 34 Sn Ag 44 Ag 45 Sn Ag 2 P Cu 6 P
Soft solders	Sn Ag 5 Sn Sb 5 Sn 50 Pb	Sn Ag 5 Sn Cu 3

[1]Derived from DVGW-Arbeitsblatt GW 2, Standard of German
Gas and Water Association.

TABLE 5

GERMAN STANDARD FOR MILD STEEL PIPES
TO BE GALVANIZED[1]

1961	1972	1978
Clean and complete welding seam	Surface must be free from scales and shells Welding seam has to be cut; maximum height is limited to 0.3 d + 0.05 d (d = thickness of pipe wall)	Mechanical tests included bending stress

[1]Derived from Standard DIN 2440.

GALVANIZED STEEL

Galvanized steel as a plumbing material is used hardly at all in the U.K. and only to a small extent in northern Europe. In contrast, central European countries commonly use galvanized steel as a plumbing material, to the extent that it accounts for 60-80 percent of all piping material.

When standards for galvanized steel were set, especially in Germany, the primary concern was to avoid corrosion due to deterioration of the material. Standards therefore had to meet noncorrosion requirements. The German Standard developed as shown in Table 5 above; the 1961 standard for the steel pipes was upgraded in 1972 to recommend cutting the welding seam, and in 1978 mechanical strength tests on the welded pipes were added to the standard. Experience had shown that the welded seams in galvanized steel pipes were the primary source of material failure due to pitting (18).

An additional standard governed the quality of the zinc layer in the galvanized steel pipes. Again, protection against corrosion (of the zinc layer) was the primary concern: "Zinkgeriesel," a type of selective corrosion of the zinc layer, leads to the presence of sandy, zinc corrosion products in house installations, often combined with heavy red water problems. Research over a period of 15 years proved that most of these problems were caused by the bad quality of the zinc layer of airblown pipes (19).

Table 6 shows the continuing development of the standard governing galvanized steel pipes. The health aspect was addressed in 1972 by reducing the cadmium level in zinc to a maximum of 0.1 percent and fixing the lead content of the zinc layer at a maximum of 1 percent. Additional recommendations on the quality of the zinc layer were introduced, and steam-blown pipes with smooth, tight zinc surfaces became standard. The most recent (1978) change to the standard occurred after research on the cadmium and lead uptake of water by corrosion of the zinc layer showed that cadmium and lead levels often were higher than drinking-water standards allowed (20), even after standing times of only a few hours. The amount of cadmium allowed in the zinc layer was therefore reduced to a maximum of 0.01 percent. The lead content of the layer, made necessary by the galvanizing process (the galvanizing tub requires a lead surface to avoid crystallization of metallic zinc), was reduced to a maximum of 0.8 percent. Lower concentrations should be reached. Data show that the mean level of 0.5 percent lead in the zinc layer of galvanized steel is not generally reached.

PLASTIC MATERIALS

The installation of plumbing systems made completely of plastic piping is only now becoming common in Europe. Many such installations have been temporary, made by introducing cross-linked polyethylene as a plumbing material. Previous attempts, mainly with polyvinyl chloride (PVC) piping

TABLE 6

GERMAN STANDARD FOR GALVANIZING
MILD STEEL PIPES[1]

1963	1972	1978
Zinc must be a minimum of 98.5% pure	Cadmium content in zinc limited to 0.1%	Zinc to be used must be free of cadmium
Coating must cover whole surface and be free from local accumulations of zinc	Lead content is limited in zinc layer to 1.0%	Cd content of the zinc is maximum of 0.01% Lead content is limited to the technological necessity: less than or equal to 0.8% Pb
Coating must have minimum quantity of 400 g/m^2	Inside surface must be free of impurities and produced by steam blowing	Production must be supervised by independent material testing institute. Pipes must be signed by durable continuous printing.

[1]Derived from Standard DIN 2444.

TABLE 7

KTW EMPFEHLUNGEN (RECOMMENDATIONS)
FOR PLASTIC AND OTHER NONMETALLIC MATERIALS
TO BE USED IN DRINKING WATER DISTRIBUTION SYSTEMS[1]

1. Material compounds (recipe) must be in accordance with
    positive lists, including limited amounts of compounds
    (each compound to be approved by toxicological working
    group)

2. General limits:

    • on taste and odor, color, turbidity and test water.
    • on migration of total organic carbon in test water.
    • on chlorine demand in chlorinated test water.

    Exposure time: 3x3 days, every 3 days change of test water

    Test water quality: Desalinated water, for Cl_2 dem. with
         0.6-0.7 ppm free chlorine

    Surface/volume ratio: Migration test - 1:1 cm^2/cm^3
                           Cl_2 demand test (depends on range
                              of use)

3. Additional requirements on special compounds to be
    migrated depending on material composition (for example,
    aromatic amines from epoxy resins, phenols from phenolic
    resins, lead from Pb-stabilized PVC pipes, etc.).

[1]Derived from recommendations given by the German
Federal Health Bureau, Berlin.

TABLE 8

MAXIMUM VALUES TO BE ALLOWED
IN KTW RECOMMENDATIONS DEPENDING
ON RANGE OF USE

RANGE OF USE	ORGANOLEPTICS (TASTE, ODOR, COLOR)	TOC mgC/m^2d	CHLORINE DEMAND $mgCl_2/m^2d$
A: Distribution pipes	Not affected S/V ratio 1:1 cm^2/cm^3	2.5	2.0
B: Reservoirs	Not affected S/V ratio 1:4 cm^2/cm^3	10	8
C: Fittings, Armatures	Not affected S/V ratio 1:6 cm^2/cm^3	15	12
D1: Gaskets with a lot of surface in contact with water	Not affected S/V ratio 1:25 cm^2/cm^3	63	75*
D2: Gaskets with little surface in contact with water	Not affected S/V ratio 1:50 cm^2/cm^3	125	150*

*In 1985, these values will change to 50 and 100.

in cold-water installations, aroused no interest except in industrial or technical water supplies where special corrosive waters (e.g., fully softened or desalinated) are distributed and most plumbing is placed on the wall.

Nevertheless, numerous nonmetallic materials are used in drinking-water distribution and plumbing installations. In addition to the hard-PVC pipes and polyethylene pipes in distribution networks, many coatings of reservoirs, fittings and armatures, gaskets, and other products in contact with drinking-water plumbing systems are made from plastics (or, more generally, nonmetallic materials).

Recommendations were developed in different European countries in order to guarantee that the quality of nonmetallic materials would have no undesirable health effects. Initial recommendations and test methods covering materials in contact with drinking water were developed (in the Netherlands and the Federal Republic of Germany) to regulate specifically the leaching of lead from lead-stabilized, hard-PVC tubes in contact with drinking water.

Later recommendations covered other questions, for example organoleptics, migration of total organic carbon (TOC), chlorine and oxygen demand, and microbiological aspects. The emphasis differs from country to country; while in Great Britain the recommendations are based mainly on microbiological aspects (21), in Germany the recommendations and methods have a more chemical and analytical direction (22).

Table 7 shows the general points of view of the German regulations (the KTW regulations). In addition to the general points covered in Table 7, the parameters vary depending not on the specific material but on the range of use. This regulation was based on the premise that the surface-area/volume ratio and the contact time negatively influence water quality. Table 8 gives the limits of the basic requirements depending on the range of use. If special requirements concerning specific compounds are found to be necessary depending on the particular use, these additional parameters will be limited with respect to the range of use of the product.

Microbiological tests are likely to be added to KTW recommendations; currently only a DVGW-Arbeitsblatt (23) covers this matter especially to describe requirements on lining materials of drinking-water reservoirs.

REFERENCES

1. Anon. 1878. Amtsblatt d. koenigl. Wuertt. Min. d. Innern. Stuttgart, No. 7, April 29, 1878: 101-106.

2. Anon. 1977. Lead in drinking water: A survey in Great Britain 1975-76. Poll. Paper No. 12, HMSO, London.

3. Gregory, R. 1981. WRC Case Studies, 2nd Seminar on Lead in Drinking Water, Paper 16 (March 3-4, 1981). Water Research Centre, Medmenham, Great Britain.

4. Elzenga, C.H.J., and A. Graveland. 1981. Studies in the Netherlands, 2nd Seminar on Lead in Drinking Water, Paper 11 (March 3-4, 1981). Water Research Centre, Medmenham, Great Britain.

5. Wagner, I., and A. Kuch. n.d. Trinkwasser und Blei, Veroeffentlichungen des Bereichs und des Lehrstuhls fuer Wasserchemie. Karlsruhe, Heft 18, ZfGW-Verlag, Frankfurt/Main.

6. Sheiham, I., and P.J. Jackson. 1981. J. Inst. Wat. Engr. Sci. 35: 491-515.

7. Schock, M.R. 1980. Journal AWWA 72: 695-704.

8. Jackson, P.J., and I. Sheiham. 1980. Calculation of lead solubility in water. TR 152, Water Research Centre, Medmenham, Great Britain.

9. Kuch, A., and I. Wagner. 1983. Water Research 17: 1303-1307.

10. Ohnesorge, F.K. 1980. Gas/Wasser 121: 515-522.

11. Department of Environment (DoE). 1982. The Glasgow duplicate diet study 1979-80. Poll. Rept. No. 11, HMSO, London.

12. Wagner, I. n.d. Verbrauchergewohnheiten - Stagnationszeiten. Auswertung von Basisdaten zum Wasserverbrauch aus unterschiedlichen Staedten der Bundesrepublik Deutschland. Unpublished.

13. Bailey, R. n.d. Influence of consumption patterns in households on lead levels in drinking water. Water Research Centre, Medmenham, Great Britain.

14. von Franque, O. 1972. Werkstoffe & Korrosion 23: 241-246.

15. Oliphant, R.J. 1982. Special Subject 18, IWSA-Congress, Zurich: 5-9.

16. Anon. 1982. Invloed van loden dienst- en binnenleidingen op het loodgehalte van drinkwater in Nederland. Report from VEWIN, KIWA, and RID. April 1982.

17. Winkler, B., and O. von Franque. 1981. Metall 35: 222-227.

18. Friehe, W. 1973. Heizung Lueftung Haustechnik 24: 51-54.

19. Werner, G., E. Wurster, and H. Sontheimer. 1973. Gas/Wasser 114: 105-117.

20. Meyer, E. 1978. La Tribune de CEBEDEAU 31: 431-441.

21. British Standards Institution (BSI). 1982. DD 82.

22. KTW-Empfehlungen des Bundesgesundheitsamtes Berlin,
     Bundesgesundheitsblatt Jarhgang 1977, 1. and 2. Mitt. ff.

23. DVGW-Arbeitsblatt W 270, Vermehrung von Mikroorganismen auf
     Materialien fuer den Trinkwasserbereich. 1983. ZfGW-Verlag,
     Frankfurt/Main.

Part V

Panel Sessions

REPORT: PANEL NO. 1

JOINING ALTERNATIVES FOR COPPER PIPE

Chairman:	Rhodes Trussell James M. Montgomery Consulting Engineers, Inc. Pasadena, California
Recorder:	Peter Karalekas EPA, Region I Boston, Massachusetts
Panelists:	A.C. Kireta Copper Development Association, Inc. Merrimack, New Hampshire
	Richard Ballentine Silver Institute Cincinnati, Ohio
	Vince Doyle Plumbing and Air Conditioning Contractors of Arizona Phoenix, Arizona
	John Courchene Seattle Water Department Seattle, Washington
	Jerome F. Smith Lead Industries Association, Inc. New York, New York
	Bill Hampshire Tin Research Institute, Inc. Columbus, Ohio
	Allen Hammer Virginia Bureau of Water Supply Engineering Richmond, Virginia
EPA Resource Person:	Marvin Gardels Water Supply Research Cincinnati, Ohio

CHARGES TO PANEL

The panel was charged with four major items for deliberation. They were:

1.   to assess extent of water quality impacts from lead/tin solder;

2.   to determine if there are acceptable alternatives to lead/tin solder;

3.   to determine if there are mitigating steps which may be taken to reduce contamination;

4.   to consider other remedial actions that could be taken by government agencies, water utilities, and plumbing and various other private groups.

To accomplish its work the panel was balanced in composition to include a water quality consultant as chairman, four representatives of solder and plumbing material suppliers (copper, silver, lead and tin), an expert in plumbing, a municipal water quality engineer, a state water supply engineer, and an EPA resource person who was expert in corrosion research. In addition there were 18 attendees representing a range of industrial, water utility, academic, and public interest groups and state and federal government agencies.

DISCUSSION

After a review of the panel charge by Chairman Trussell, the seven panelists spoke briefly in turn on various subjects related to the panel charges. A.C. Kireta of the Copper Development Association (CDA) felt that any problem was confined to soft acidic waters. In a CDA study of 142 buildings in Connecticut ranging in age from zero to 120 years old, only two had lead concentrations greater than the maximum contaminant level (MCL) for lead. Mr. Kireta recommended that all new plumbing systems be flushed before being placed in service and questioned present sampling methods for lead in tap water. Among the alternatives to lead/tin solder is tin/silver, which Kireta felt was a good alternative. Brazing of pipe was not desirable because of the many fittings now being manufactured with plastic internal parts which would not withstand high brazing temperatures. Tin/antimony is also an acceptable alternative; however, its use on potable water lines would be difficult to enforce because of the problem of switching by plumbers who would have lead/tin on hand for waste piping. Kireta felt that water treatment to reduce corrosion was the major solution to solving any existing water quality problems.

Richard Ballentine, representing the Silver Institute, pointed out that 95-5 tin/antimony was not a good wetting solder whereas tin/silver was very good and had a higher strength. He estimated that the cost for plumbing a house with tin/silver solder was only $5 to $10 more than tin/lead because of reduced labor costs and the need for less solder due to shorter overlap in couplings.

Vince Doyle, of the Plumbing and Air Conditioning Contractors of Arizona, felt that the solution to the problem with lead/tin solder was to upgrade training procedures for plumbers. He showed examples of poorly soldered joints with excessive flux and solder on the interior of the pipe as well as perfectly soldered joints with only a very thin line of solder exposed to water. Mr. Doyle felt there would be no problem if all joints were soldered efficiently. He also felt that tin/antimony was more difficult to work with.

John Courchene, of the Seattle Water Department, discussed the corrosion problems that had occurred in Seattle. Studies had shown that 60 percent of water samples had exceeded the MCL for iron and 13 percent for lead. Among the problems associated with corrosion were high plumbing repair bills ($7.5 million/year), staining, and health. The solution developed for Seattle involved raising the pH of water from 6.8 to 8.3, encouraging the use of noncorrosive materials such as plastic, and banning the use of solder containing lead. These measures have resulted in the reduction of lead, copper, cadmium, zinc, and iron in Seattle water.

Jerome Smith, Lead Industries Association, conceded that high lead concentrations could be found in the first sample drawn from a faucet in the morning. However, he questioned the meaning of the sample and urged that epidemiological studies be done to determine if lead/tin solder had any effect on blood lead levels. Among the solutions recommended by Smith were treating water to reduce its corrosivity, chemical flushing of all newly plumbed water lines, and the use of preforms in the joints of copper tubing to prevent solder from flowing into the pipe itself which would result in its retention in the joint.

Bill Hampshire, Tin Research Institute, pointed out that lead from soldered joints in tin cans and from gasoline had been reduced significantly. In studies by the Tin Research Institute there are a number of solders that can replace tin/lead. Among the more promising are a tin/copper solder which is weaker than lead/tin but still strong enough for its intended use; other solders under development are stronger.

Allen Hammer, Virginia State Health Department, said that some communities still had lead service lines and that lead had been found there in excess of 10 times the MCL. He described a situation occurring in June 1983 in a house where 18 of 31 samples were greater than the MCL. The piping was removed and replaced using tin/antimony solder. Further sampling found no samples containing lead greater than the MCL. In another house, the highest lead concentration found was 6.1 mg/l before flushing the water. Hammer felt there was no predictability of lead results. Among the actions taken by Virginia were a news release recommending flushing of water lines before consumption, appointment of a task force on building code changes, and further study of the problem. In preliminary results of this new study 11 of 65 samples had lead greater than the MCL, with ranges from 0.001 to 0.4 mg/l lead in waters that ranged from pH 6.0 to 9.1. In another study by a private lab only 1 of 105 samples contained lead greater than the MCL.

In the discussion period following the presentations by panel members, a number of questions and issues were raised. Opinions differed, from one individual who felt there was no problem to others who advocated a total ban on solder containing lead.

A total ban would deal with any future problems, but a number of individuals pointed out the necessity of dealing with the current problem. It was suggested that water utilities be required to treat their water to reduce its corrosivity toward plumbing. The observation was made that there is no standardized technique for sampling for lead. Regarding the issue of the occurrence of the problem being confined to soft water areas, Schock pointed out that in studies in Illinois, the problem of lead was identified in communities with very hard, high-pH water, indicating a more widespread problem. Regarding the ability of manufacturers to meet the demand if lead/tin solder were banned, one individual felt that the marketplace would meet any demand that developed.

FINDINGS AND CONCLUSIONS

1. Lead is present in some waters due to corrosion of solder. The problem has occurred in both hard and soft water areas.

2. New installations, which also may contain particles of lead solder and flux, and supplies with soft, acidic water, may be the worst-case situations.

3. Alternative joining materials such as tin/silver or tin/antimony solder are available should lead/tin solder be displaced.

4. Water treatment methods are available to reduce the problem in certain areas. Also, properly soldered joints, chemical flushing, and preforms may be useful in reducing lead in drinking water.

5. Proper plumbing practice achieved through communication with plumbing groups and training can have a role in reducing the problem.

RECOMMENDATIONS

Although a number of recommendations were put forth by various panel members and the audience, it was generally felt a consensus was reached on the following areas.

1. More study is needed on the presence of lead from corrosion of solder in a variety of water qualities.

2. A standardized sample collection method should be developed to be used in these studies and in monitoring of public water supply systems.

3. Additional studies are needed on treatment techniques to reduce corrosion of lead from solder since simple pH adjustment may not be appropriate for all waters.

REPORT: PANEL NO. 2

METAL PIPE AND FITTING ALTERNATIVES FOR PLUMBING

Chairman: J. Edward Singley
                James M. Montgomery
                   Consulting Engineers, Inc.
                Gainesville, Florida

Recorder: Frank A. Bell, Jr.
                Office of Drinking Water
                Washington, D.C.

Panelists: James R. Boates
                Boston Plumbing Company
                Natick, Massachusetts

                Dodd S. Carr
                International Lead Zinc
                   Research Organization, Inc.
                New York, New York

                Anthony Colucci
                Colucci and Associates
                Morgan Hill, California

                Larry Galowin
                National Bureau of Standards
                Washington, D.C.

                Joe Glicker
                Bureau of Water Works
                Portland, Oregon

                Stanley Mruk
                Plastic Pipe Institute
                New York, New York

                Robert A. Wilson
                Copper Development Association
                Purcellville, Virginia

CHARGES TO PANEL

Charges to the panel were to assess the extent of water quality
impacts from metal plumbing pipe and fittings and to consider mitigating
steps such as water quality control, use of alternative grounding
electrodes, limitation or control of some materials, and possible standards
and certification programs. Possible training and educational materials
development or teaching programs were also to be considered.

To accomplish its charges the panel was balanced in composition and
included a water quality/corrosion expert as chairman with discussants
including industrial representatives in the areas of copper, galvanized
steel, and plastic materials and metal fittings; a plumbing contractor; a
municipal water quality engineer; and a government researcher in plumbing
systems. In addition there were 15 attendees from a range of industrial,
consulting, and municipal and state government sources.

DISCUSSION

Metal pipe and fitting alternatives were outlined by the discussants;
key points were as follows:

- Copper pipe is the dominant household plumbing material with
  principally esthetic effects from copper leaching and staining.

- Galvanized steel pipe has had a level rate of use in the United
  States and is commonly used in many communities. Leaching effects
  are principally esthetic with elevated zinc levels, although traces
  of lead and cadmium may be present.

- Lead pipe and fittings are seldom used except in Chicago where the
  plumbing code requires that service lines be lead pipe. (See "Lead
  Product Utilization Survey of Public Water Supply Distribution
  Systems throughout the United States," April 1984, which is
  included as Appendix A to this panel report.) Lead
  levels are elevated in water standing in lead pipes, more so in
  low-pH, low-alkalinity waters but also in a variety of other waters.

- Metal fittings may also serve as a source of leached contamination,
  for example lead, where the subject material is contained in the
  fitting.

Water quality control was urged as the main solution to control
corrosion in all types of metal pipe by the metal industry
representatives. In general they urged that local utilities provide
corrosion control treatment and that decisions regarding piping materials
be made at the local level on a case-by-case basis.

On the other hand the water utility representative said that plumbing materials, where identified as the source of contaminants, should be controlled by plumbing code modification. He also discussed the utilization of EPA's Drinking Water Advisory #73 in pursuit of appropriate corrosion control sampling and inquired what national results had been found from the 1980 corrosion monitoring regulations. While not a part of this panel discussion, attention was frequently turned to the subject of lead/tin solder. An undercurrent of concern regarding recent and ongoing moves to ban lead/tin solder was evident through the whole panel session.

A lack of communication was cited by the plumbing representative, who was unaware of the water quality impacts of plumbing materials. He emphatically stated that plumbers wanted to safely and efficiently transmit drinking water to the consumer in the same quality received from the public utility, but he was unaware of some of the problems that had been mentioned.

Other needs for research and development were highlighted in the areas of:

- standardized sampling and analytical procedures for the study of leached contamination.

- improved means of studying corrosion scientifically and efficiently.

- the problem of corrosion from grounding and stray electric current and the effective use of dielectric or separating fittings between dissimilar materials.

FINDINGS AND CONCLUSIONS

While this was a stimulating panel session, the divergent and conflicting views and time limitations prevented a full coverage of all subjects. For the subjects covered, the chief conclusion appeared to be that lead pipe contributes some lead contamination even in waters considered noncorrosive; its use in new plumbing installations has been abandoned in nearly all jurisdictions; there is no logical requirement for its use to continue. The majority conclusion was that its future use should be discouraged or banned.

Panel consensus was evident on the general subject of communication. It was agreed that more formal communication on plumbing materials and water quality needs to be provided, particularly to the plumbing and code development and enforcement communities.

Regarding other materials or problems, there was insufficient information or time to cover the assigned subjects. The questions at hand were important and should be covered in subsequent meetings in more depth.

APPENDIX TO PANEL SESSION 2.

LEAD PRODUCT USE SURVEY OF PUBLIC WATER SUPPLY DISTRIBUTION SYSTEMS
THROUGHOUT THE UNITED STATES

INTRODUCTION AND SUMMARY

A survey was conducted by each of the 10 EPA Regional Water Supply
Branches to assess the extent of lead utilization in water distribution
systems throughout the United States. A total of 153 public water systems,
selected from 41 states, including the District of Columbia and Puerto
Rico, were examined.

In a previous survey that was conducted almost 60 years ago ("Action
of Water on Service Pipes" by Wellington Donaldson, Journal AWWA, 11:649
(1924), 48 percent of the 539 cities that were evaluated had utilized lead
service pipes. The results of the present survey confirm that lead
products still exist in many water distribution systems and that the
utilization of such products in the past was widespread.

With regard to lead in drinking water, many studies show that lead is
seldom found in measurable concentrations in water sources. Lead, however,
is found frequently in corrosive tap water. Corrosive water may be reduced
by a number of measures, such as adjusting the pH, etc.

The problem of corrosion of lead has become such an important issue
that the EPA has proposed interim regulations requiring the determination
of the presence of specific materials in water distribution systems and
monitoring for characteristics of corrosivity of the water. Furthermore,
recent studies have indicated lead-tin solder can impart excessive lead
concentrations to drinking water.

ACKNOWLEDGEMENT

This survey was prepared and reviewed by David Chin and Peter
Karalekas of the U.S. EPA Region I Water Supply Branch staff. Special
thanks go to Terry Regan for performing the tedious task of typing this
survey. Region I would like to commend the efforts of Water Supply Branch
members from the other nine regions who contributed the data and
information for this survey and to thank Frank Bell for his encouragement
and assistance.

METHODOLOGY

The survey consisted of conducting telephone interviews or sending out
the questionnaires to state counterparts or to the particular water
systems. Most of the systems that were surveyed represented those that

served large populations (91 out of the 153 systems served populations exceeding 100,000 people).

The main problem with this survey was the reliability of the responses that were obtained. A number of questions that were asked involved making rough estimates of the use of lead products in water distribution systems. In some of the systems that we evaluated, recordkeeping just did not exist in the past. Some water system operators or superintendents were more experienced and knowledgeable than others, which in turn resulted in more reliable estimates. Also, some superintendents may just be more cooperative than others, and would be willing to divulge more information and make an effort to do some research. It is difficult to take into account a number of these subjective factors.

In order to obtain uniform response from each of the ten regions, the following questionnaire was developed and suggested to be utilized in this survey:

- What is the total population served by the water system?

- What is the total number of services in the water system?

- What service line materials are being used for new installations?

- What part of the service line does the water system own?

- Were lead or lead-lined service pipes ever installed in the distribution system? If so, how many still remain in the distribution system and where are they located?

- Were lead gooseneck service connections ever used? If so, how many still remain in the distribution system and where are they located?

- Were lead sweatjoints ever used? If so, how many still remain in the distribution system and where are they located? Are they still being used?

- What other service line materials have been used in the past?

- Does the water system have an ongoing replacement program for lead materials utilized in the distribution system?

- Has the system had any problems with high lead concentrations in the distribution system?

- What is the pH of the water in the distribution system? Has the pH been adjusted higher to prevent corrosion problems?

OBSERVATIONS

We primarily focused our attention on the information obtained on the water systems' utilization of lead services, lead goosenecks and lead sweatjoints. The following table summarizes our findings:

Type of Response	# of Systems That Had Utilized Lead Services (Out of 153 Responses)	# of Systems That Had Used Lead Goosenecks (Out of 152 Responses)
Affirmative	112 (73.2%)	94 (61.8%)
Unknown	5 (3.3%)	15 (9.9%)
None Known	36 (23.5%)	43 (28.3%)
Not Asked or No Response	—	1

Approximately 73 percent of the systems we surveyed indicated that lead service lines had been installed in the past. In fact, the City of Chicago still installs lead services for those locations utilizing a size pipe of under and including 2". Lead services are still part of the accepted plumbing code in Chicago.

Lead goosenecks have also been widely used. Nearly 62 percent of the water systems had utilized lead goosenecks. Approximately half of the water systems questioned reported the use of lead sweatjoint in their systems. It is assumed that, because of the universal use of lead-tin solder for joining copper pipes, lead in solder would be found in all of the systems surveyed.

As for the responses regarding the ownership of the service line, they varied widely from the water supplier owning the service line from the main to the meter, property line, or curb, to the consumer owning the entire service line. The significance of these responses is that for many water systems, water suppliers have no responsibility over the utilization of lead products in the home plumbing area.

Few water suppliers have active ongoing replacement programs for any lead materials that have been installed in the distribution system. It appears that water suppliers will only replace a lead service line or a lead gooseneck, for example, when a leak exists.

Many system reported pH values of less than the optimum for controlling lead corrosion. Yet, nearly all of the responses we received on the question of whether there is a problem with high lead concentrations in the distribution system were negative. Once-per-year monitoring for lead required by the NIPDWR would certainly account for many of the negative responses.

We have summarized, alphabetically by state, the information that was submitted from all of the ten EPA Regions in the following table.

Name of Utility and/or Public Water Supply Location	Estimated # of people served	Total # of services	Service line material used for old installations	Service line material used for new installations	Use of lead services % or # still left	Use of lead goosenecks % or # still left	Use of lead sweatjoints % or # still left	pH in the distribution system and if adjusted
Birmingham, Alabama	650,000	150,000	galvanized iron, PVC	copper	none known	none known	none known	8.5 lime added
Montgomery, Alabama	175,000	54,232	galvanized iron, lead	copper	# unknown, replaced if found	none known	Yes, all have been replaced	8.9 lime added
Fayetteville, Arkansas	45,000	11,765	galvanized iron	copper, PVC	none known, replaced if found	none known	Yes, 3-4%	7.5 no adjustment
Fort Smith, Arkansas	71,629	29,000	copper, galvinized iron	copper, PVC, polyethylene	# unknown, general area known	Yes, # unknown	Yes, # unknown	7.5 no adjustment
Hot Springs, Arkansas	43,000	17,700	PVC, copper, galvanized	copper	Yes, none remain	Yes, < 200	none known	8.2 no adjustment
Pine Bluff, Arkansas	60,000	20,077	copper, PVC	PVC	unknown	unknown	unknown	7.1 no adjustment
El Centro, California	26,402	6,100	copper, lead	copper	300	unknown	unknown	7.4 - 7.6
San Diego, California	929,000	208,946	lead used until 1940	plastic, copper	10,000	10,000	10,000	7.6 - 8.2
Arvada, Colorado	86,409	24,500	copper, galvanized	PVC	Yes, # unknown	Yes, # unknown	# unknown, still being used	7.4 adjusted
Aurora, Colorado	187,000	45,000	polyethylene	copper	almost all have been replaced	Yes, 12	none known	7.5 no adjustment
Denver, Colorado	1,000,000	225,000	copper	copper	Yes, # unknown	Yes, # unknown	Yes, # unknown	7.2 - 7.8 adjustment at one plant

Name of Utility or Public Water Supply	Estimated # of people served	Total # of services	Service line material used for old installations	Service line material used for new installations	Use of lead services % or # still left	Use of lead goosenecks % or # still left	Use of lead sweatjoints % or # still left	pH in the distribution system and if adjusted
Bridgeport Hydraulic Co., Bridgeport, CT	336,185	90,000	brass, galvanized iron	PVC, copper	1%	2%	still being used for services	6.8 - 7.5
Hartford MDC, Hartford, CT	391,000	87,000	N/A	copper, brass	none known	Yes, # unknown	Yes 1500	7.01
New Britain Water Department New Britain, CT	99,302	15,802	N/A	copper	Yes, # unknown	1,500	still in use	8.2 - 8.3
New Haven Water Department New Haven, CT	371,259	90,000	brass, lead, galvinized iron	copper, PVC ductile iron	most have been removed	30,000	N/A	7.2
Stamford Water Company Stamford, CT	84,000	19,714	brass, lead, galvinized iron	copper	24	532	N/A	6.7 - 7.1
Waterbury Water Department Waterbury, CT	124,000	25,000	N/A	copper, ductile iron	none known	Yes 10,000	N/A	7.2
Dover Water Department Dover, Delaware	27,500	7,428	N/A	copper	none known	50%	N/A	8.0 - 8.2
Artesian Water Co. Newark, Delaware	140,000	36,449	N/A	copper, PVC	none known	4%	N/A	7.3 - 7.7
Newark Water Department Newark, Delaware	30,000	6,800	N/A	copper	none known	none known	N/A	6.7 - 7.2
Wilmington Water Department Wilmington, DE	140,000	34,252	N/A	copper	<5%	none known	N/A	7.0 - 7.7
Wilmington, DE Suburban Water Co. of Claymont, DE	75,000	23,000	N/A	polyethylene	<1%	<1%	N/A	7.1 - 7.9
District of Columbia	756,500	120,000	N/A	copper	60,000	none known	N/A	7.6 - 8.3

Name of Utility or Public Water Supply	Estimated # of people served	Total # of services	Service line material used for old installations	Service line material used for new installations	Use of lead services % or # still left	Use of lead goosenecks % or # still left	Use of lead sweatjoints % or # still left	pH in the distribution system and if adjusted
Orlando, Florida	300,000	76,100	galvanized, copper	PVC, copper	Yes, all have been replaced	none known	none known	7.8 no adjustment
Tallahassee, Florida	100,000	42,000	galvanized, PVC, lead	PVC, copper	# unknown, some being replaced	# unknown, some being replaced	none known	8.1 adjusted with caustic soda
Atlanta, Georgia	650,000	127,636	galvanized, PVC	copper	# unknown	none known	none known	7.3 phosphate and lime added
Savannah, Georgia	150,000	45,000	cast iron, steel	polyethylene or polybutylene	Yes, # unknown	Yes, # unknown	none known	7.0 - 7.3 no adjustment
Boise, Idaho	105,000	38,482	code accepted	code accepted	100 known	100 known	Yes 4,350	7.1 - 7.7
Cour d'Arlene, Idaho	21,000	9,000	galvinized iron, copper, kalmain	polyethylene, PVC	none known	200	none found	7.0 - 7.2
Idaho Falls, Idaho	41,000	15,600	code accepted	code accepted	none known	10,000	20,000	7.2 - 7.3
Lewiston, Idaho	17,000	4,900	code accepted	copper	# unknown, very early use	not known	# unknown, very early use	7.4 - 7.6 adjusted
Twin Falls, Idaho	26,000	9,489	galvanized	plastic, code accepted	none known	Yes, # unknown	none	7.9 adjusted
Chicago, Illinois (city proper only)	3,000,000	400,000	ductile iron, lead	lead for 2" & under; ductile iron for >2"	all pipe 2" and under remain	For all pipe 2" and under	?	8.1 - 8.5 adjusted
Decatur, Illinois	94,000	31,000	lead, galvinized iron	copper	<100 replaced when found	<100 replaced when found	?	9.5
Gary - Hobart Water Corp. Gary, Indiana	100,000 (Gary Only)	30,000	lead until 1950, copper	copper	25,000 replaced when found	25,000 replaced when found	?	7.6

Name of Utility or Public Water Supply	Estimated # of people served	Total # of services	Service line material used for old installations	Service line material used for new installations	Use of lead services % or # still left	Use of lead goosenecks % or # still left	Use of lead sweatjoints % or # still left	pH in the distribution system and if adjusted
Des Moines, Iowa Water Works	253,078	68,301	galvanized, lead	copper, ductile iron	Yes, # unknown	Yes, # unknown	Yes, # unknown	9.3
Dubuque, Iowa Water Works	62,321	20,390	lead used prior to 1928	copper, ductile iron	5,000	not known	Yes, # unknown	9.5 - 9.6
Sioux City, Iowa Utilities Dept.	82,000	25,000	N/A	copper, PVC, ductile and cast iron	12,000	Yes, # unknown	2,000	7.2 - 7.5
Kansas City, Kansas Board of Public Utilities	160,000	56,000	galvanized	copper, PVC	a few, # unknown	several hundred	several hundred	7.8 - 8.1
Johnson County, Kansas Water District #1	205,000	65,972	galvanized and cast iron	copper	none known	none known	none known	9.0
Topeka, Kansas Water Department	140,000	45,000	galvanized wrought iron	copper, ductile iron, brass	135	none known	none known	9.2 - 9.5
Wichita, Kansas Dept. of Water and Water Poll. Control	280,000	115,000	lead prior to early 1940's	PVC, copper	<15,000 older parts of city	100	none known	8.2
Lexington, Kentucky	290,000	70,000	galvanized, lead, copper PVC	PVC	# unknown, replacement program	# unknown, replacement program	none known	7.2 phosphate and lime added
Louisville, Kentucky	670,000 - 720,000	202,000	galvanized	copper	Yes, # unknown	none known	none known	8.2, lime added
Portland, Maine Water District	141,240	38,000	galvanized iron	copper, ductile iron	none known	200	N/A	6.6 - 6.7
Annapolis, Maryland DPW	.95,000	75,000	N/A	N/A	none known	none known	N/A	8.7
Baltimore, Maryland	1,100,000	368,000	N/A	N/A	<0.1%	<0.1%	N/A	7.8

Name of Utility or Public Water Supply	Estimated # of people served	Total # of services	Service line material used for old installations	Service line material used for new installations	Use of lead services % or # still left	Use of lead goosenecks % or # still left	Use of lead sweatjoints % or # still left	pH in the distribution system and if adjusted
Cumberland, Maryland	40,000	17,200	N/A	N/A	none known	none known	N/A	7.4
Hagerstown, Maryland	70,000	16,892	N/A	N/A	none known	none known	N/A	7.5
Washington Suburban Sanitary Commission Maryland	1,705,000	N/A	N/A	N/A	none known	none known	N/A	7.7
Boston, MA Water and Sewer Commission	600,000	90,000	cast iron, Lead	copper, ductile iron, cement-lined	50%	50%	N/A	8.8
Brockton, MA Water Department	89,040	21,455	N/A	copper	none known	<10% 2,000	N/A	9.6 - 9.8
Cambridge, MA Water Department	100,361	13,250	N/A	copper	<10%	<10%	N/A	8.6 - 8.8
Fall River, MA Water Department	100,000	15,500	N/A	PVC, copper	10%	10	N/A	8.5
Lawrence, MA Water Department	62,000	12,000	N/A	copper	1500	none known	N/A	8.0
Lowell, MA Water Department	94,000	18,000	lead, copper, tin alloy	copper	<25% 4,500	<25% 4,500	N/A	7.2 - 7.3
Lynn, MA Water Department	80,000	17,800	copper, lead, cement-lined, galvinized	copper	20% 3,500	40% 7,000	N/A	6.5 - 7.5
New Bedford, MA Water Department	100,169	22,000	N/A	copper	30% 6,600	30% 6,600	N/A	7.0
Newton, MA Water Department	91,066	25,612	N/A	copper, PVC, galvanized, cement-lined	1,000	4,000 - 5,000	N/A	8.8

Name of Utility or Public Water Supply	Estimated # of people served	Total # of services	Service line material used for old installations	Service line material used for new installations	Use of lead services % or # still left	Use of lead goosenecks % or # still left	Use of lead sweatjoints % or # still left	pH in the distribution system and if adjusted
Quincy, MA Water Department	91,200	21,000	N/A	copper	<200	Yes, # unknown	N/A	8.8
Somerville, MA Water Department	80,600	14,000	N/A	copper	<20% 100-200 replaced/yr	Yes, # unknown	N/A	8.8
Springfield, MA Water Department	241,500	43,381	galvanized iron	copper	none known	unknown	N/A	6.8
Worcester, MA Water Department	172,342	37,000	N/A	copper, ductile iron	unknown	unknown	N/A	6.5 - 6.6
Kalamazoo, Michigan	130,000	35,000	lead	copper	5%	5%	?	7.0 no adjustment
Rochester, Minnesota	60,000	17,500	lead, copper, galvanized	copper	30%	30%	?	7.4 no adjustment
Jackson, Mississippi	210,000	66,000	PVC, galvanized	copper, brass	Yes, all have been replaced	Yes, all have been replaced	none known	8.9 - 9.0 lime added
Meridian, Mississippi	49,000	16,000	galvanized	copper	Yes, # unknown	none known	none known	8.5 lime added
Cape Girardeau, Missouri Mo. Utilities Co.	35,000	12,332	galvanized	polyethylene, copper	Yes, none remain	unknown	none known	8.3
Columbia, Missouri, Water and Light Department	62,000	17,000	plastic	copper	# unknown, old section of town	# unknown, old section of town	none known	8.5
Florissant, Missouri Water Department	70,000	15,000	N/A	copper, brass, ductile and cast iron	50	none known	not in use since 1944, may still exist	9.6 - 9.8
Independence, Missouri Mo. Water Company	120,000	40,000	galvanized	copper, ductile iron	very few exist	2,000	none known	9.8

Name of Utility or Public Water Supply	Estimated # of people served	Total # of services	Service line material used for old installations	Service line material used for new installations	Use of lead services % or # still left	Use of lead goosenecks % or # still left	Use of lead sweatjoints % or # still left	pH in the distribution system and if adjusted
Jefferson City, Missouri, Capital City Water Company	33,000	9,300	galvanized, cast iron	copper	Yes, # unknown	Yes, # unknown	Yes, # unknown	9.7 - 9.9
Joplin, Missouri Water Works Co.	55,000	17,000	copper, cast iron, tubeloy	polyethylene tubing	1,500	none known	none known	7.5
Kirkwood, Missouri Water Department	27,987	9,761	galvanized iron	copper	Yes, # unknown	Yes, # unknown	none known	10.0
Lee's Summit, Missouri Water Company	38,000	10,500	galvanized	copper	30-40 %	unknown	Yes, # unknown	7.8 - 8.3
Raytown, Missouri Water Company	15,000	6,500	galvanized	copper	none known	Very few, if any	Very few, if any	9.9
St. Joseph, Missouri Water Company	91,518	30,213	iron tubeloy, plastic, A-C, lead	PVC, copper	8779	8779	17,558	7.6
St. Louis, Missouri Water Division	455,000	118,500	galvanized	copper, ductile iron	unknown as to number & location	Yes, # unknown	still being used	8.4 - 10.2
St. Louis County, Missouri Water Company	1,000,000	260,910	cast iron, galvanized, A-C	copper, PVC, ductile iron	28,000	Yes, # unknown	Yes, # unknown	9.6 - 9.8
Sedalia, Missouri Water Department	22,000	8,833	galvanized	copper, ductile and cast iron	Yes, # unknown	Yes, # unknown	Yes, # unknown	8.0 - 8.4
Grand Island, Nebraska Utilities Dept.	33,944	10,739	cast and ductile iron	copper	2,000	none known	none known	7.3
Lincoln, Nebraska Water Department	174,560	52,488	galvanized, cast iron	copper, ductile iron	innumerable	innumerable	innumerable	7.5
Omaha, Nebraska Utilities District	431,000	126,234	galvanized iron, tube alloy	copper, brass, ductile iron	25,000 - 30,000	25,000 - 30,000	wiped-lead joints	9.0

Name of Utility or Public Water Supply	Estimated # of people served	Total # of services	Service line material used for old installations	Service line material used for new installations	Use of lead services % or # still left	Use of lead goosenecks % or # still left	Use of lead sweatjoints % or # still left	pH in the distribution system and if adjusted
Manchester, New Hampshire Water Works	100,000	22,000	galvanized iron	copper	none known	250	none known	6.6 - 7.8
Albuquerque, New Mexico	342,000	104,400	galvanized	polystyrene, copper, cast iron	1,000 locations known	none known	none known	8.1 No adjustment
Sante Fe, New Mexico	50,000	14,000	galvanized	cast or ductile iron, copper <2"	none known	1,000 locations known	several miles of old lines	7.8 adjusted
Elizabethtown, New Jersey Water Company	500,000	150,080	N/A	PVC for <2", copper, ductile iron	<5% very few	<1% very few	N/A	6.8 - 7.6
Hackensack, New Jersey Water Company	800,000	180,709	N/A	copper for <2", ductile iron	<5% 150 replaced/yr	25,000 100 replaced/yr	N/A	7.8
Monmouth, NJ Consolidated Water Company	249,000	70,121	N/A	copper, PVC, ductile iron	2,972	unknown	N/A	7.5 - 7.8
Newark, New Jersey Water Department	380,000	58,000	N/A	copper, brass, ductile iron	about 75%	none known	N/A	7.5 - 7.9
Passaic Valley, New Jersey Water Commission	287,316	63,000	N/A	copper, ductile iron	about 25%	unknown	N/A	7.1 - 7.3
Buffalo, New York Division of Water	358,000	90,000	N/A	copper, ductile iron, PVC	# unknown	same	N/A	7.5 - 7.8
Erie, County, New York Water Authority	317,000	99,148	N/A	copper, ductile iron, PVC	# unknown	same	N/A	7.8
Jamaica, New York Water Supply Co.	650,000	118,000	N/A	copper, ductile iron	# unknown, many	# unknown, many	N/A	6.0 - 6.9
New York City, New York Aquaduct System	7,071,000	820,000	N/A	copper, brass ductile and cast iron	# unknown, many	# unknown, many	N/A	6.3 - 7.4

Name of Utility or Public Water Supply	Estimated # of people served	Total # of services	Service line material used for old installations	Service line material used for new installations	Use of lead services % or # still left	Use of lead goosenecks % or # still left	Use of lead sweatjoints % or # still left	pH in the distribution system and if adjusted
Suffolk County, New York Water Authority	850,000	246,898	N/A	copper, ductile iron	<1%	none known	N/A	6.7-7.2
Charlotte, North Carolina	350,000	106,342	galvanized, black iron, PVC	copper	Yes, all have been replaced	Yes, all have been replaced	none known	9.3 lime added
Raleigh, North Carolina	165,000	47,000	galvanized iron	copper, brass	none known	Yes, all have been replaced	none known	7.2 NaOH & phosphate added
Bismarck, North Dakota	35,000	8,881	cast iron	copper, plastic	none known	none known	none known	9.3
Fargo, North Dakota	61,281	25,159	iron, lead, copper	copper	Yes 2,500	# unknown, used a long time ago	Yes, 2,500	9.0 - 9.5 lime added
Grand Forks, North Dakota	53,060	8,927	black iron	copper	Yes 2,000	none known	Yes, 4,000	8.6 - 9.2 lime added
Lawton, Oklahoma	100,000	40,000	galvanized	copper	Yes, 250 remain in older areas	Yes, 150 remain in older areas	none known	7.6
Oklahoma City, Oklahoma	460,000	165,000	N/A	copper	Very few remain in older areas	Yes, in old sections of town	Yes, in old sections of town	10.5
Tulsa, Oklahoma	420,000	140,000	galvanized, copper, PVC	copper, plastic, brass, cast iron	Yes, very few remain	Yes, very few remain	none known	7.5
Corvallis, Oregon	43,000	12,000	galvanized	3/4" meter uses 1"copper or plastic	none known	Yes, all known removed	none known	N/A
Eugene, Oregon	103,000	34,777	polyethylene, copper, polybutylene	Copper	Yes, # unknown	Yes, # unknown	Yes	7.2 - 7.3 adjusted
Hillsboro, Oregon	35,000	10,000	galvanized, copper, polybutylene	code accepted	none known	none known	none known	6.9 adjusted

Name of Utility or Public Water Supply	Estimated # of people served	Total # of services	Service line material used for old installations	Service line material used for new installations	Use of lead services % or # still left	Use of lead goosenecks % or # still left	Use of lead sweatjoints % or # still left	pH in the distribution system and if adjusted
Portland, Oregon	650,000	120,000	galvanized	copper is used up to the meter	Yes, # unknown	Yes, 10,000	Yes 20,000	6.8 No adjustment
Salem, Oregon	120,000	45,000	code accepted	code accepted	unknown	none found	some, # unknown, still in use	low 6's No adjustment
Erie, Pennsylvania Bureau of Water	205,000	52,000	copper, galvanized iron	Type k copper	none known	2% remain from 1920's service area	used until early 1960's	7.3 - 7.6
Philadelphia, Pennsylvania Water Department	1,685,000	522,000	copper, galvanized & black iron	copper	Yes, # unknown	Yes, # unknown	leadite joints used; most replaced	7.0 - 8.5
Philadelphia, PA Suburban Water Company	902,000	287,000	N/A	copper	<200	none known	N/A	7.0 - 7.6
Pittsburgh, Pennsylvania Water Department	424,000	89,000	copper	copper	30%	none known	Yes, 5% remain	7.6 - 7.8
Western Pennsylvania Water Company of Pittsburgh, PA	500,000	126,512	galvanized, cast and wrought iron	copper and ductile iron	Yes 5,318	Yes, # unknown	Yes, # unknown	7.1
Metropolitano A San Juan, Puerto Rico	711,999	178,000	N/A	copper, PVC, ductile iron	none known	none known	N/A	7.4 - 8.2
Enrique Ortega La Plata, Puerto Rico	363,936	90,984	N/A	copper, PVC, ductile iron	none known	none known	N/A	7.4 - 8.0
Ponce Urbano, Puerto Rico	241,540	60,385	N/A	copper, PVC, ductile iron	none known	none known	N/A	7.4 - 8.0
Aqua Dilla, Puerto Rico	114,497	28,624	N/A	copper, PVC, ductile iron	none known	none known	N/A	7.8 - 8.2
Caguas, Puerto Rico	109,236	27,309	N/A	copper, PVC, ductile iron	none known	none known	N/A	7.2 - 7.6

Name of Utility or Public Water Supply	Estimated # of people served	Total # of services	Service line material used for old installations	Service line material used for new installations	Use of lead services % or # still left	Use of lead goosenecks % or # still left	Use of lead sweatjoints % or # still left	pH in the distribution system and if adjusted
Pawtucket, Rhode Island Water Department	108,750	21,000	brass, galvanized & cast iron	copper	lead-lined 10% 2,100	none known	N/A	7.2 – 7.3
Providence, Rhode Island Water Department	283,816	69,121	N/A	copper and ductile iron	25%	unknown	N/A	10.0
Warwick, Rhode Island Water Department	78,500	26,000	N/A	copper	100	none known	N/A	10.0
Charleston, South Carolina	350,000	60,000	galvanized steel	copper	Yes, # unknown	Yes, # unknown	Yes, # unknown	8.0 – 8.3
Columbia, South Carolina	231,700	70,440	galvanized	Type k copper	none known	none known	none known	6.9 adjusted
Huron, South Dakota Water Department	13,000	3,500– 4,000	lead and galvanized	copper	Yes, # & location unknown	possibly, # & location unknown	possibly, # & location unknown	8.8 – 9.2 adjusted
Pierre, South Dakota	11,973	3,800	ductile and cast iron	PVC	Yes # unknown little used	Possibly, # & location unknown	Possibly, # & location unknown	7.5 – 8.2 no adjustment
Sioux Falls, South Dakota	81,343	25,000	galvanized	copper	Yes, # & location unknown	Yes, # & location unknown	Yes, # & location unknown	7.0 – 8.0 no adjustment
Memphis, Tennessee	800,000	221,709	PVC, galvanized	copper	Yes, some remain in older areas	none known	none known	7.2 phosphate added
Nashville, Tennessee	350,000	105,000	cast iron, galvanized, PVC	copper	# unknown, in old parts of the city	none known	Yes, # unknown	7.8 – 8.0 caustic soda added
Logan, Utah	28,000	56,000	galvanized and ductile iron	copper with brass/bronze fittings	Yes, a little used	Yes, a lot used	Yes, some used with goosenecks	N/A
Odgen, Utah	64,000	20,000	N/A	copper	none known	Yes, many years ago # unknown	none to curb; many inside bldgs	7.1 – 7.2

Name of Utility or Public Water Supply	Estimated # of people served	Total # of services	Service line material used for old installations	Service line material used for new installations	Use of lead services % or # still left	Use of lead goosenecks % or # still left	Use of lead sweatjoints % or # still left	pH in the distribution system and if adjusted
Provo, Utah	75,000	15,000	galvanized	copper	Yes, a few remain	unknown	none known	8.4 – 8.5
Arlington County, Virginia	165,000	34,000	N/A	copper	200	3,000	N/A	8.0
Fairfax County, Virginia Water Authority	650,000	134,000	N/A	copper	none known	none known	N/A	7.0+
Henrico County, Virginia	170,000	48,200	N/A	plastic, copper	none known	1%	N/A	7.3 – 7.9
Newport News, Virginia	333,000	90,000	N/A	copper	<2%	<2%	N/A	7.1
Norfolk, Virginia	270,000	61,700	N/A	PVC	3%	N/A	N/A	7.2
Richmond, Virginia	225,000	62,500	N/A	copper	25,000	none known	N/A	7.2 – 7.6
Bellingham, Washington	50,000	17,000	galvanized	copper or plastic	not known	Yes	Yes	6.8 – 6.9 adjusted
Everett, Washington	300,000	90,000	galvanized and copper	anything Code allows	believed most has been removed	probably	none known	7.4 Adjusted
Seattle, Washington	1,115,000	168,371	galvanized, copper, plastic, kalmain	Code Accepted	Yes, # unknown	unknown	probably	8.00 soda ash and lime added
Spokane, Washington	180,000	62,000	copper, galvanized	Type F copper galvanized,	If used all have been replaced	If used all have been replaced	may have	7.8 none
Tacoma, Washington	214,000	68,362	galvanized, copper	plastic, galvanized copper	none known	none known	Yes, a lot still in use	6.8 – 7.2

REPORT: PANEL NO. 3

PLASTIC PIPE AND FITTINGS

Chairman: Gary Englund
Minnesota Division of Environmental Health
Minneapolis, Minnesota

Recorder: Harry Von Huben
EPA, Region V
Chicago, Illinois

Panelists: Alan J. Olson
B.F. Goodrich Co.
Cleveland, Ohio

David Spath
California State Department of Health
Berkeley, California

Tom Adams
Adams, Broadwell and Russell
San Mateo, California

Nina McClelland
National Sanitation Foundation
Ann Arbor, Michigan

Ray Lee
American Water Works Service Co., Inc.
Hadden Heights, New Jersey

Daniel W. Hurley
Associated Lead, Inc.
Philadelphia, Pennsylvania

EPA Resource   Ray Jones
  Persons   EPA, Region X
Seattle, Washington

Koge Suto
EPA, Region III
Philadelphia, Pennsylvania

CHARGES TO PANEL

The panel was charged with assessing the extent of impact to drinking water quality due to leaching from plastic pipe, permeation of plastic pipe, and leaching from joining compounds and solvents; further charges were to assess possible responses to the leaching and permeation problems. Other approaches to the issue of public health and the use of plastic pipe were also considered.

To accomplish its work the panel was balanced in composition to include a state water supply engineer as chairman; representatives of various industrial (plastics and metals), plumbing, and third-party standards interests; water utility and State health engineers; and two knowledgeable EPA technologists as resource persons. In addition there were 23 attendees representing a range of water utility, industrial, research, and state and federal governmental interests.

<u>Charge No. 1</u>:   Assess the extent of impact to drinking water quality due to leaching from various types of plastic pipe (PVC, CPVC, PB, etc.).

Discussion

(Discussion on this item was strictly limited to leaching from the pipe itself, with the problems of joining materials considered as a separate issue.)

The argument was made by the plastic pipe interests that plastic pipe has been used and tested for over 20 years with no indication of adverse health effects due to leaching. The point was also made that plastic pipe is regularly tested by the National Sanitation Foundation (NSF) with no adverse health risk being noted.

The counter was made that the report being prepared by SRI International may indicate significant leaching problems. The point was also made that NSF works only from established health limits. The example was given that if chloroform at 99 g/1 was found in a sample of pipe, it would be "approved" by NSF — yet, most authorities would agree that the level should be lower if possible. It was also noted that NSF carefully looks at the analysis of the pipe material, but does not necessarily try to determine other materials that may be present due to the manufacture of the pipe.

Although everyone, more or less, had to concede that additional adverse health effects due to pipe material leaching <u>could</u> be found in the future, the plastic pipe interests were reluctant to endorse a general statement that more research should be done on possible leaching problems.

Panel Opinion

     Majority opinion: The majority did not feel there is a problem of
adverse water quality due to leaching from plastic pipe, based on known
health effects information.

     Minority opinion: The minority felt that there is a potential problem
and more information and research is needed. They particularly felt there
is a greater potential problem with some types of plastic pipe than with
others.

Charge No. 2: Assess the extent of impact to drinking water quality due to
               permeation of plastic pipe.

Discussion

     The point was made by plastic industry interests that a great amount
of plastic pipe is in use and, percentage-wise, the instances of detection
of permeation are extremely small. It was also stated that most
authorities will agree that it is not good practice to install any pipe in
contaminated soil. Neither of these points were contested.

     But, the counter point was made that most pipe is installed when the
soil is "clean," with the expectation that it will remain clean for many
years, and then, in some instances, the soil becomes contaminated later,
creating a problem if the pipe is permeable.

     Plastic industry interests stated that they do not deny there is a
permeation problem. They said they want to "get a handle on the problem"
and "want to be the first to make recommendations on specific areas where
plastic pipe should not be used." They also stated that they are looking
at protection measures that possibly could be incorporated into manufacture
or added during the installation of the pipe.

     The point was made that a number of jurisdictions at a local level had
already taken steps to control locations where permeable pipe may be used.

Panel Opinion

     It was unanimously agreed that there is a demonstrated susceptibility
of plastic pipe to permeation by organic chemicals. However, it was
recognized that the susceptibility is somewhat variable for different types
of plastic.

<u>Charge No. 3</u>:   Assess the extent of impact to drinking water quality due to leaching from joining compounds, solvents and other materials.

Discussion

There was general agreement that there is some degree of leaching of compounds and solvents used in pipe joining. There appears to be little information at present on the degree and duration of the problem.

There were opposing views on whether the leaching poses a potential public health risk. One faction stated that there is no data demonstrating a health risk, that the amount of leached material is, over all, very small, and that the problem only persists for a few years after the pipe is installed. Others felt that the exposure is significant enough to seriously investigate the potential health danger.

There was also discussion on the potential of greater-than-necessary leaching of joining compounds due to what is perceived as widespread use of incorrect or "multi-purpose" joining compounds. According to plastic industry representatives, multi-purpose compounds should be used only when pipes of different types are joined; otherwise, only the material specified for the particular type of pipe should be used.

Panel Opinion

Majority opinion: The majority agreed there is a problem of leaching from joining compounds and more information is needed on the health risks that this creates.

Minority opinion: The minority felt there is some leaching that takes place from joining compounds, but that it is not of public health significance.

<u>Charge No. 4</u>:   Assess possible responses to leaching problems from piping, joining compounds, and other materials:

- Limit the use of some materials.

- Improve installation techniques.

- Strengthen third-party standards and monitoring programs.

- No change.

Discussion

It was generally agreed that, although there is concern, we are nowhere near considering prohibiting the use of any materials for general

use. It was also mentioned that any "limiting" of use in the future might
be in regard to ingredients, rather than particular products.

The fact that California and various local jurisdictions have already
imposed some regulation on use demonstrates that there is a definite need
to strengthen third-party standards and model codes.

Panel Opinion

There was agreement that there is a definite need to strengthen
third-party agreements and monitoring to review state, local, and model
codes and to continue to review and improve installation techniques. There
is a continuing need, as an industry, to review the compounds and solvents
used in the manufacture and installation of the products. There is also a
definite need for improved information, communications, and instruction
from the manufacturers, regulators and trade groups.

<u>Charge No. 5</u>:   Assess possible responses to the problems of permeation and
water quality:

- Limit the use of plastic pipe.

- Develop new non-permeable pipe materials.

- No change.

Discussion

It was generally agreed that there is a need to notify users and the
public that plastic pipe should not be used in contaminated soil.

There was considerable interest in the California regulation
prohibiting use of plastic pipe under conditions considered to pose a
possible adverse health effect. The regulation is not well known or well
enforced, and it was written several years before permeation problems were
understood, but it is of interest as a start in recognizing and addressing
the problem.

It was agreed that there is a great need for the plastic pipe
industry, trade groups, the plumbing industry, and others to bring about an
awareness of the impacts of possible permeation.

The plastic industry representatives stated that they are working on
development of non-permeable pipe materials. There are also new
technologies being developed for improved installation (including soil
barriers), preventing migration, etc., that might also be applicable in
preventing permeation of plastic pipe.

It was unanimously agreed that "no change" in addressing the problems
of permeation was not appropriate.

Panel Opinion

It was agreed that:

- There are areas where there is a need to limit the use of plastic
  pipe because of existing or potential ground or environmental
  contamination.

- There is a need for the industry or other bodies to inform
  potential users of plastic pipe of the limitations of the products.

- There is a need for further investigation into the development of
  non-permeable pipe or other means of preventing pipe permeation.

<u>Charge No. 6</u>: Detail any other approaches to coping with problems relating
                 to public health and use of plastic pipe.

Discussion

It was again discussed that there is a definite need to positively
identify the various types of pipe material and joining compounds or
solvents, so that only "compatible" materials will be used. Color coding
or other means could be used. This would make it easier for the user to
determine what products are appropriate and it would make it much easier
for the field inspector to assure that correct materials are being used.

It was stated that NSF is doing some testing for leaching and
permeation, and indications are that this program will be expedited in view
of the remarks and discussions at this meeting.

There is a great need for the manufacturers and trade associations to
"get the word out" to the users and installers on any problems, special
conditions of use, etc.

It was suggested that possibly a clause could be incorporated into the
NSF or other certification, stating that use of incorrect joining compounds
would nullify the certification.

Panel Opinion

It was agreed that:

- There is a need for better communications between the industry, the
  trades, and consumers and installers on potential problems and/or
  locations where plastic pipe should not be used.

- There appears to be a need for better identification (color coding,
  labeling) of plastic pipe and joining compounds/solvents to better
  assure the use of the proper joining material for each type of pipe.

- "Universal" solvents are supposed to be used only for joining different types of pipe, but are apparently being indiscriminately and generally used due to incomplete information furnished by the suppliers, or because the trades feel they are the easiest to use.

<u>Charge No. 7</u>: What are the indicated desirable or recommended remedial actions by governmental, manufacturing, trade, code, or testing groups?

Discussion

There was unanimous agreement that there has not been adequate communications from any of the groups regarding the plastic pipe problems that have been discussed at this meeting.

It was also discussed that there is a real need for additional testing and research, particularly in regard to "real-world" conditions of installation as they may differ from laboratory-controlled conditions.

There was also mention of a need for consistency in the information put out by the various interested groups. It was also stated that it is often difficult for someone with problems to find a source of information. A "hotline" number for plastic pipe would be very helpful in this regard.

The matter was also discussed, without resolution, that there is a need for each of the interested groups to be doing something specific toward investigating the problems that have been discussed.

Panel Opinion

It was agreed that there is a need to improve communications between all of the interested groups to get consistency in the information being generated.

OTHER PANEL OBSERVATIONS AND RECOMMENDATIONS

A representative of the cast-iron pipe industry stated that gaskets are available for cast-iron pipe that are used when the pipe is to convey organic compounds, so the same material should be usable to prevent contamination of water in the pipe if the soil is contaminated. It seems that this is an important point if permeation of regular gaskets is considered to be a problem, and should be made readily available to the trade and public.

Although it was not specifically discussed by the panel, it should be recognized that the "problems" of plastic pipe actually sort out into three different "jurisdictional" areas:

1. The water mains and water services, generally to the service
    stop, are usually of a material approved by and installed in a
    manner specified by the water utility.

2. Some jurisdictions allow the installation of other material past
    curb stop, so the material could be different. The problems of
    permeation through pipe is primarily a problem with these
    jurisdictions.

3. The third problem the panel felt may be significant is from
    joining material or solvents. This is primarily a problem of
    interior piping, which is under the jurisdiction and inspection
    (if any) by the local building inspector.

Somehow, it will be necessary to communicate with each of these
jurisdictions regarding the problems that will affect them.

REPORT: PANEL NO. 4

REGULATORY AND COMPLIANCE ASPECTS

Chairman:	Raymond Jarema Connecticut Department of Health Services Hartford, Connecticut
Recorder:	Charles D. Larson EPA, Region I Boston, Massachusetts
Panelists:	James Sargent Wisconsin Bureau of Plumbing Sun Prairie, Wisconsin
	Richard B. Howell, III Delaware Bureau of Environmental Health Dover, Delaware
	Donald Dickerson Dickerson Associates Van Nuys, California
	Charles Buescher, Jr. Continental Water Co. St. Louis, Missouri
	Hugh F. Hanson Consumers/Pirnie Utility Services Co. White Plains, New York
EPA Resource Persons:	Peter Lassovszky and Arthur Perler Office of Drinking Water Washington, DC

CHARGES TO PANEL

The charges to the panel were:

1. Examine the means by which materials/contamination problems can
   be controlled through Maximum Contaminant Level (MCL) and

monitoring requirements (either existing or as potentially modified) of the Primary Drinking Water Regulations, as regards:

- Lead pipe and fittings.

- Copper pipe, joints, and fittings.

- Galvanized pipe and fittings.

- Plastic pipe, joining compounds, and fittings.

2. Examine the means by which materials/contamination problems can be controlled through:

- State and local building or plumbing codes.

- Regional or national model codes.

3. Examine monitoring aspects of code compliance. (For example, how would compliance with a code/solder requirement be monitored?)

4. Examine other mechanisms for regulatory or compliance change.

5. Assess any indicated desirable or recommended remedial actions by:

- Federal and state drinking water agencies relating to changes in drinking water regulations, monitoring, and implementation.

- Regional or national model code groups.

- Local and state code groups.

To accomplish its work the panel was balanced in composition to include a state water supply engineer as chairman, a state code official, two engineers knowledgeable in water utility operations, a former chairman of a national model code development committee, a state water supply engineer involved in changing his state's plumbing code, and two EPA resource persons who are knowledgeable in corrosion and additive aspects. In addition there were 20 attendees representing a range of water utility, industry, research and state and federal governmental interests.

DISCUSSION

Panel members as well as most of the attendees of this panel session felt the approach to solving regulatory and compliance problems must be multifaceted. Limiting the toxic elements in materials used for plumbing as well as treating water to reduce corrosiveness are both essential. Because of the short time allotted for the panel to come up with

recommendations, the best approach appeared to be to define
responsibilities and assign those responsibilities to the most appropriate
organizational entities.

The U.S. Environmental Protection Agency (EPA) was assigned the
responsibility of defining the problem, assessing or compiling existing
data on a regional or national basis, developing primarily MCLs, and
monitoring requirements to regulate corrosion byproducts. It was also
thought that EPA should be responsible for preparing source documents for
the states. It was recommended that the EPA conduct a national sampling
survey similar to the National Organics Reconnaissance Survey (NORS) or
National Organics Monitoring Survey (NOMS) surveys. Before undertaking the
survey there should be a standardization of sampling techniques and a study
designed which would incorporate uniformity in sampling. It was also
considered very critical that educational materials for states and local
government authorities should be developed by EPA. Water Supply Guidance
No. 73 (WSG 73) was passed out to those attending the panel discussion.
WSG 73 specifically deals with monitoring and sampling techniques to be
used for evaluating the effects of corrosion. There was concern voiced
that WSG 73 was not and had not been distributed widely enough and it
should certainly be made available to utilities. Also WSG 73 may serve as
the basis of a standardized sampling program. A copy of WSG 73 is included
as Appendix A to this panel report.

The state's responsibility was defined as being the enforcement agent
over the water purveyor. It was also felt that a model code should be
developed at the state level. However, there was some debate over whether
the model code should be at the state or at the federal level. There was
some consensus among the group that communication was necessary to improve
state agency interaction and coordination. This interaction includes water
purveyors, health departments, and building and housing departments at the
state level.

The water purveyor's responsibility was defined as the provision of
safe drinking water to its consumers. Water safety is assured through a
regular monitoring program; therefore, it is the water purveyor who will be
collecting the data in a uniform manner. It is the water purveyor who
implements treatment to reduce corrosivity. It is also the water
purveyor's responsibility to transmit educational material on an ongoing
basis to the consumer, thereby helping to educate the public.

Manufacturers of materials, including materials used to make pipe and
joints, should reassess standards on an ongoing basis. Third-party review,
for example by the National Sanitation Foundation, was discussed as being a
fairly good mechanism for a central standard-setting process.

At the local government level, the consensus was that the
responsibility for enforcement of plumbing as well as building code
requirements should be administered at this level. Because the inspection
and approval process takes place at this level, enforcement should also be
administered at this level.

Consumers also have responsibilities and they should use the educational materials distributed to them. The consumer-water purveyor relationship is an essential one that must have credibility and trust incorporated into it.

A building owner's responsibility is to comply with existing codes, standards, and regulations that pertain to his facilities, including building, plumbing, and health codes. Building owners have an ultimate responsibility to their clients in meeting compliance requirements by ensuring that public safety is addressed.

The responsibility for research was not designated, although the general consensus was that additional epidemiological and health effects research is needed.

In summary, communication and public education must be immediately developed and implemented. Corrosive water should be made less corrosive, and exposure to toxic compounds in plumbing materials should be reduced as much as possible. EPA should define the corrosion problem and develop primary MCL and monitoring requirements to regulate hazardous corrosion byproducts. EPA should develop a standardized sampling procedure to help identify distribution system corrosion problems. A national survey of corrosion byproducts using the standardized sampling procedure is recommended.

APPENDIX TO PANEL SESSION 4

UNITED STATES ENVIRONMENTAL PROTECTION AGENCY

DATE: AUG 4   1982

SUBJECT: Water Supply Guidance Number 73

FROM: Victor J. Kimm, Director
Office of Drinking Water (WH-550)

TO: Water Supply Representatives, Region I-X
Holders of Water Supply Guidance Series

Attached is Water Supply Guidance Number 73 for Monitoring an Sampling Techniques to Determine Corrosion Products, includi lead, in water supply distribution systems.

Several Regional Offices have requested clarification and
guidance on sampling techniques to be used for compliance
monitoring for metals in drinking water, and on the applicability of MCLs to corrosivity and corrosion products in water.

Neither the National Interim Primary Drinking Water Regulations
(NIPDWR), nor the Manual for Interim Certification of Laboratories address the issue of sampling techniques for metals.
By definition, an MCL is the maximum permissible level of a
contaminant in the water delivered to the consumer. Conventional sampling techniques, in which lines are flushed in an
effort to provide a sample representative of the water provided by the supplier of water, is usually considered to be
appropriate in terms of the MCL definition. The NIPDWR contained minimum requirements for the MCL, monitoring and reporting, with the expectation that States and other authorities
would adapt them to meet local needs. In the cases of substances that are primarily introduced into drinking water during
transit, it is probable that more extensive site specific
monitoring and sampling methods should be chosen by the local
or State authority to obtain a better indication of actual
human exposure from drinking water.

Paragraph 141.42 of the NIPDWR requires that community water
suppliers identify specific materials of construction that are
present in their distribution systems. However, additional
monitoring requirements were not specified to determine the
presence of the corrosion products that are contributed from
these materials to the drinking water. Corrosivity of the
water is to be determined on the basis of the Langelier Index,
and in certain cases, the Aggressive Index. When there is a
potential for an MCL to be exceeded, additional actions are
obviously warranted.

Although the regulations are not explicit regarding sampling
techniques for corrosion products in water, the States obviously have the authority to require specified monitoring in
instances where it is felt that the water quality indices are
inadequate to determine the corrosive tendency of the water or
where regulated contaminants are potentially present. In such
instances, the attached guidance produced by the Criteria and
Standards Division can provide a practical tool to determine
the representative range of specific contaminant levels to
which the consumer might be exposed as a result of the interaction of the water and the materials of construction in the
distribution system. The sampling techniques described are
not federally enforceable, but are intended only as a guidance.

This approach was adopted from the methods that have been successfully used by Region I in dealing with lead contamination resulting
from aggressive water in contact with lead pipe and services.

Water Supply Guidance Number 73

Guidance for Monitoring and Sampling Techniques to Determine Corrosion Products, Including Lead, in Water Supply Distribution Systems

Introduction

The presence of contaminants in the water at the consumer's tap is a function of both treated water quality and distribution system factors that can lead to localized contaminant problems. Because the occurrence of corrosion related contaminants in the water distribution system are often localized, random sampling techniques are not always successful in detecting their presence. Thus, the minimum monitoring requirements as prescribed by the NIPDWR in 1975 may not be appropriate in some situations to realistically determine the extent of consumer exposure to contaminants resulting from corrosive water.

The 1980 amendments to the NIPDWR's established a one year corrosivity monitoring program which was intended to cause systems to identify circumstances where corrosion related contamination may occur, and to encourage appropriate corrective actions. Corrosivity measurements required under these regulations are determined indirectly using the indices (see 40 CFR 141.42(c)) which are based on calcium carbonate saturation.

The indices specified in the 1980 amendments generally measure the tendency of the specific water to form a protective calcium carbonate ($CaCO_3$) layer. The index value is dependent on the pH, alkalinity, hardness, total dissolved solids (TDS) and the temperature of the water. Each of these parameters may independently affect the corrosive tendencies of the water so that waters may be corrosive even though the index indicates non-corrosive conditions.

Therefore, it is generally agreed that the indices may only be applicable to determine the corrosive tendencies of the water within narrow pH ranges, and then only if sufficient amounts of calcium and alkalinity are present in the water. These indices do not take into account the use of corrosion inhibiting chemicals such as hexametaphosphate and bimetalic phosphates. Additionally, calcium carbonate stability, as well as other parameters, may change in the distribution system because of retention in distribution reservoirs and rechlorination.

The sampling techniques suggested herein provide a method to monitor and verify the effective corrosivity of the water and the effectiveness of treatment to control corrosion for those

instances where the water quality indices are not applicable.
For additional details involving the applicability of corrosion
indices and monitoring, refer to the Statement of Basis and
Purpose for the August 27, 1980, Amendments to the National
Interim Drinking Water Regulations.

Purpose

This Guidance provides a suggested protocol for water suppliers
who desire to determine if corrosion products are present in
those situations where the existing regulations may not provide
sufficient specificity. Specifically, the sampling techniques
herein can help to determine the range of specific contaminant
concentrations which a consumer might experience as a result of
the interaction of the water and the materials comprising the
water delivery system. In addition, the Guidance provides
criteria to determine when additional actions might be appropriate. The Agency realizes that, due to large variations in
water quality and distribution materials in any water supply
system, implementation of sampling protocol should be done on
a case-by-case basis. The protocol below is a suggested model
that States and utilities may use in its entirety or in part
as determined by specific situations.

Guidance for the Selection of Appropriate Sample Sites

Sample sites for monitoring should be established at locations
where the potential for consumer exposure to a specific contaminant would be greatest. For instance, monitoring for lead
should be done at sample sites in the distribution system where
the presence of lead pipes or service has been verified. The
rationale for problem specific site selection as opposed to random
sampling, is that corrosive by-products (i.e., lead, zinc, copper
and others) are not normally found in distribution waters in high
concentrations except where they are being dissolved from piping
materials due to corrosion. Table 1 provides information on the
type of contaminants associated with various delivery system construction materials. Table 2 contains a suggested ratio of sampling
sites to population served. The design of an appropriate sampling
regimen also needs to reflect local conditions including hydraulic
configuration, water aggressiveness, sampling costs, analytical
costs and other factors. In instances where the presence of a
material in the distribution is known to be localized, the ratio
of sample sites to population served should be adjusted to reflect
the portion of the distribution system affected, rather than the
whole water supply system. In addition, sections of the distribution system that are served by multiple sources of different water
quality should be treated as a separate entity for sampling purposes
to account for variations in water quality in the affected portions
served by the various sources.

Table 1

Recommendations for Sampling Parameters

The following parameters are suggested for analysis for different types of pipe material:

Pipe Material	Parameter of Interest
Ductile and Cast Iron (Unlined)	Iron
Copper	Copper, Lead from Solder Joints
Lead	Lead
Galvanized Steel	Zinc, Cadmium, Lead, Iron
Copper Alloys	Copper, Zinc, Lead

Note: Asbestos fibers from asbestos cement pipe was not included as a suggested parameter. Monitoring asbestos cement pipe deterioration is best done in the distribution mains rather than by the sampling program suggested in this guidance.

Table 2

Recommended Number of Sampling Sites to be Monitored on a Quarterly Basis for the First Year

Population Served	Number of Sampling Sites
<100	1
101 - 1000	2
1001 - 5000	5
5001 - 10,000	7
>10,001	10

Note: It is expected that localized variables affecting corrosion are more numerous in larger distribution systems that in smaller ones. For this reason, proportionately more sample sites are recommended for larger water systems in order to better account for these variables.

The water purveyor should use discretion in determining the
need for monitoring for specific contaminants. Where previous
records and experience has indicated that the water is not
corrosive to a specific material or that certain piping materials
have not been used, the purveyor should modify his sampling and
analytical program to reflect that information.

Once the number and type of samples to be analyzed are established, the water supplier can rank the location of points to
be sampled using the following suggested order of priority:

1.   Locations where corrosion problems are suspected due to
      consumer complaints or other indications,

2.   Buildings with interior plumbing containing specific
      materials of interest such as lead, galvanized steel,
      copper, etc.,

3.   Service lines containing specific materials of interest
      such as lead, galvanized steel, copper, etc.,

4.   Recently constructed buildings served by the utility, and

5.   Several randomly selected control sites throughout the
      distribution system.

For example, in lead monitoring, a water supplier would sample
for lead at locations meeting priority "1" first. If not enough
buildings are available with appropriate interior plumbing, he
would determine the remaining locations using priority "2", "3",
"4" and/or "5". Sample sites to monitor other contaminants such as
iron, copper, zinc and cadmium should be established in a
similar manner. The suggested order of priority was established
on the basis of maximum potential exposure to the consumer.
Priorities "1", "2" and "3" are self explanatory. Priority "4"
was established on the basis of field data which indicates that
corrosion rates are much higher in the newly constructed plumbing
than older ones.

There are a number of options available for the utilities to
perform a survey to locate and identify specific sample sites.
Consultation of records and archives may be one of the best
approaches for some larger and older systems to determine the
general location and presence of contamination sources. This
approach will reduce manpower and time requirements in the
field to pinpoint the exact location of the source of contamination. Often, the local or county health department might have
already conducted a survey involving health-related contaminants
in the water as a result of distribution system corrosion.

An alternative approach is to have utility personnel, such as meter readers or meter repairmen develop information as a routine part of their normal jobs. It is usually possible to determine the type of material used for service lines by checking the connections at the meter. Another method which could be utilized is to mail questionnaires. However, it should be noted that such an approach usually produces poor results, since only a small portion of the customers may reply, and the information provided may not be reliable. The surveys could be conducted by initiating a direct house-to-house search for the identity, type and source of materials in the distribution system. This latter approach would only be attractive for small utilities having few connections.

<u>Guidance for Collection of Samples for Analysis*</u>

To determine the representative range of contaminant levels to which the consumer is exposed during the course of daily activities, the samples should be collected early in the morning. The samples should be drawn from interior faucets to get water that has been in intimate and sustained contact with the interior plumbing, the service line and water main as follows:

Sample 1 - The sample reflecting interior plumbing condition is collected immediately upon opening the faucet. This represents water that usually has been standing in the fixture and interior plumbing for a 6 to 8 hour period.

Sample 2 - The sample reflecting service line condition is collected after the water has been running and the sample collector feels the water temperature change from warm to cold. Since water would be expected to warm slightly after standing in interior plumbing, this cold water represents water that had been standing just outside the foundation of the house in contact with the underground service line for the same 6 to 8 hour period.

Sample 3 - The sample representative of the water main is collected after allowing the water to run for at least three minutes after the initial temperature change was noted. This water would have had a minimum contact time with the service line and interior plumbing and probably represents the ambient distribution water quality. The length and size of pipes in the building should be considered to determine if the 3 minute contact criterion is applicable.

---

*Karalekas, P. C., Jr., Ryan, C. R., Larson, C., Taylor, F., "Alternative Methods for Controlling the Corrosion of the Lead Pipe." Paper presented at the Annual Conference of the New England Water Works Association, Boston, Massachusetts, 1977.

This method of sample collection provides an indirect profile
of consumer exposure to corrosion related contaminants. Water
that has been stationary in the interior plumbing overnight has
had maximum contact time with the source of contamination and
represents water that may be drunk by the consumer early in
the morning. The second sample represents the water that was
standing in the service line and may represent water used for
preparation of breakfast beverages such as coffee or tea. The
third sample, taken after running the water, represents water
that is commonly drawn during the same day when water is more
frequently used. It is obvious that no single sampling protocol
can represent every scenario of exposure. The suggested protocol is an attempt to estimate typical exposure potential. Utilities and States are encouraged to use other protocols if a specific
exposure situation can be sampled with more confidence. Further
study is required before a more accurate universal protocol can
be articulated.

Recommendations for Sampling Frequency and Interpretation of
Results

To account for seasonal variations in water quality affecting
corrosion, the sampling should be repeated at least quarterly
for a period of one year. This sampling frequency may be reduced
in instances where the water obtained is solely from the
ground and it has been established that the water quality is
constant through the year. In reviewing these quarterly samples
and in making determinations about corrective action, the water
supplier should consider any first draw sample or any average
value (average of 3 samples taken at one location) that are
greater than the ambient or background levels for the particular
contaminant being evaluated. Neither the first draw sample
nor the average values of the 3 samples taken at each location
would constitute an MCL violation per se, but they would give
good indications of the possibility that an MCL is being exceeded
for some portion of the time and that a part of the population
is being exposed to elevated levels of the corrosion product.
It may be that a reassessment of the MCL compliance monitoring
regimen should be undertaken to assure that the monitoring program
being used is indicative of the quality of water being consumed
in those parts of the system that are being affected by corrosive
waters.

If the above sampling results show that the highest average
concentration from any one set of the four seasonal samples
at each sampling point is below MCL values, reducing the frequency of sampling at each site to one sample per year is justi-
fied, but this sample should be obtained during the seasonal
period that revealed the greatest corrosion induced contamination. If the above sampling survey results show that the
average concentration of contaminants from any one set of the

four seasonal samples at any sampling point exceeds the MCL
(or other level of concern for unregulated contaminants) remedial action should be contemplated. Such actions might
include treatment additions or modifications to reduce the
corrosivity of the water, the use of corrosion inhibiting
chemicals, or changes to piping materials to alleviate the
health risk associated with isolated local problem areas.
Quarterly sampling should continue until the average concentration of contaminants is known to be consistently below the
MCL (or other level of concern for unregulated contaminants.)

A maximum contaminant level is defined to be, "...The maximum
permissible level of contaminant in water which is delivered to
the free flowing outlet of the ultimate user of a public water
system...contaminants added to the water under circumstances
controlled by the user, except those resulting from corrosion
of piping and plumbing caused by water quality, are excluded
from this definition." MCLs have been established for lead and
cadmium in the National Interim Primary Drinking Water Regulations, and for copper, zinc and iron in the National Secondary
Drinking Water Regulations. These MCLs define the level of
contaminant at which corrective actions are appropriate and
required. However, the wording of the Safe Drinking Water Act
and the legislative history that accompanied the Act direct
that any human exposure to potentially harmful contaminants
should be minimized to the extent feasible. Concentrations
of corrosion by-products such as lead and cadmium can reach
levels in excess of the MCL in very short periods of time (40
minutes or less in some studies).* Therefore, even though the
average values obtained using the previously discussed sampling
protocol or samples taken in accordance with the minimum requirements set out in the National Interim Primary Drinking Water
Regulations do not technically exceed the MCL, any elevated
contaminant levels indicate the potential for excessive exposures
and prudent public health policy would dictate corrective or
remedial actions.

---

*Peacock, Stuart J., "Factors Influencing Household Water Lead:
A British National Survey". Archives of Environmental Health,
Vol. 35, No. 1, Jan-Feb 1980, pp. 45-51.

Briton, A., "Factors Influencing Plumbosoluency in Scotland",
Journal of the Institution of Water Engineers and Scientists,
Vol. 35, July 1981, pp. 350-364.

REPORT: PANEL SESSIONS

SUMMARY OF CONCLUSIONS AND RECOMMENDATIONS

Frank Bell

Office of Drinking Water
U.S. EPA
401 M Street, S.W.
Washington, D.C. 20460

BACKGROUND

The panel sessions of the seminar were divided into four subject areas:

1. Joining Alternatives for Copper Pipe

2. Metal Pipe and Fitting Alternatives for Plumbing

3. Plastic Pipe and Fittings

4. Regulatory and Compliance Aspects.

Charges to each panel followed the general framework of assessing the current state of knowledge, identifying problems, assessing alternative solutions, and recommending future steps to cope with the identified problems.

KEY CONCLUSIONS AND RECOMMENDATIONS

The following conclusions and recommendations represent a sense of the meeting in the important areas of concern, especially emanating from the panel discussions.

LEAD/TIN SOLDER

Lead is present in drinking waters, both hard and soft, due to corrosion of lead/tin solder. New installations, which also may contain particles of lead solder and flux, and supplies with soft, acidic water may be the worst-case situations.

Several state and local jurisdictions have acted and others are
considering acting to ban lead/tin solder from potable water plumbing.
There were strong differing opinions and no consensus could be reached
regarding the desirability of banning the future use of lead/tin solder.
Alternative joining materials such as tin/silver or tin/antimony solder are
available should lead/tin solder be replaced.

Water treatment methods are available to reduce the problem in certain
areas. Since poor plumbing practice appears to be involved in several
incidents of excess potable water lead levels, improvements in plumbing
practice through communication and training can have a role in reducing the
problem. In addition the use of chemical flushing and preforms may be
useful in reducing lead contamination levels.

<u>Recommendations</u>:

    a. While the occurrence of high lead levels attributable to lead/tin
       solder was accepted, it was recommended that more study and
       analysis of data were required to assess the extent and severity of
       lead contamination in a variety of water qualities.

    b. Additional investigative and research work were also recommended to
       develop a standardized sampling method and to refine water quality
       treatment and control measures.

LEAD PIPE AND FITTINGS

The use of lead pipe for potable water has been abandoned by most
jurisdictions throughout the United States with the exception of the City
of Chicago, where the plumbing code requires that service lines of 2 inches
or less in diameter be lead pipe. Lead levels are elevated in water
standing in lead
pipes, more so in low-pH, low-alkalinity waters but also in a variety of
other waters. Since there are economic alternative materials, including
copper, galvanized steel, and plastic (currently used in many cities as
shown in an EPA survey of April 1984), there is no logical requirement for
the continued use of lead pipe and fittings.

<u>Recommendation</u>: The majority recommendation was that the future use of
lead pipe and fittings should be discouraged or banned.

TREATMENT OR WATER QUALITY ADJUSTMENT

A basic paper was presented on techniques of water treatment to
minimize corrosion. In the panels on joining alternatives for copper pipe
and on metal pipe and fittings, a repeated theme emphasized the problems of
controlling leaching from existing systems. Corrosive waters will increase
the rate of leaching from lead pipes, a residual of which exists in many
communities, and from copper pipe joined by lead/tin solder. Excess

contaminant leaching has been associated with a variety of water
qualities. Since simple pH adjustment may not be appropriate for all
waters, more study and research are needed on treatment techniques to
reduce corrosion.

Recommendations:

   a. All public water utilities should review their water quality and
       potential corrosion problems and should take steps to make their
       water quality less corrosive where elevated contaminant levels
       occur.

   b. Additional study and research should be conducted to define the
       occurrence of corrosion in a spectrum of water qualities and to
       develop appropriate anticorrosion treatment methods.

MONITORING FOR CORROSION BYPRODUCTS

     Several panels pointed out questions with respect to corrosion
sampling techniques and the need for developing a standardized approach.
When EPA's Water Supply Guidance 73 (WSG 73) was passed out at one panel
meeting, concern was expressed that it had not been distributed widely
enough and was virtually unknown in many communities. EPA also promulgated
corrosion monitoring regulations in 1980 and their results remain to be
determined.

     EPA was assigned responsibility for defining the corrosion problem and
for developing primary Maximum Contaminant Levels (MCLs) and monitoring
requirements to regulate corrosion byproducts. The state's responsibility
was defined as being the enforcement agent over the water purveyor. The
water purveyor's responsibilities were defined to include implementing
corrosion monitoring in a uniform manner, adjusting water quality to
minimize corrosion, and rendering a safe quality water to consumers.

Recommendations: EPA should define the corrosion problem and develop
primary MCL and monitoring requirements to regulate corrosion byproducts.
Development of a standardized sampling procedure that would help identify
distribution system corrosion problems and a national survey of corrosion
byproducts utilizing the standardized sampling procedure are recommended.

PLASTIC PIPE AND FITTINGS/LEACHING

     The leaching of toxic substances from plastic pipe to drinking water
has been the subject of long-term standards and monitoring under the
National Sanitation Foundation (NSF) third-party program. Few problems and
no adverse health risk have been noted in over 20 years of surveillance for
covered products. However, not all byproducts of plastic manufacture may
have been covered in the NSF program: a minority felt the need for further
research, particularly with respect to some materials more than others.

Contamination of drinking water from the use of solvents and joining compounds in the field assemblage of some plastic pipes represents a significant concern. There was also discussion on the potential of greater than necessary leaching from the misuse of incorrect or "multipurpose" compounds. The multipurpose compound would be used when two pipes of different materials are joined; otherwise, only the material specified for a particular type of pipe would be used. Apparently a lack of materials identification and incomplete use instructions from suppliers may be responsible for much of the misuse of multipurpose joining compounds.

Recommendations: There was majority agreement that there is a problem of leaching from joining compounds and that more information is needed on the health risks that this creates. Specific recommendations were to strengthen third-party agreements and monitoring; to review state, local, and model codes; and to review and improve installation techniques. The industry was urged to review the compounds and solvents used in the manufacture and installation of plastic products.

PLASTIC PIPE AND FITTINGS/PERMEATION

Research has shown that plastic pipe and fittings are permeable to organic chemicals such as gasoline and trichloroethylene and that drinking water contamination has occurred in several instances where such chemicals have come into contact with plastic pipe. Research has also shown that asbestos/cement pipe and the gaskets used for joining ductile iron pipe are permeable to some organic chemicals.

While the extent of contamination from the permeation of plastic pipe appears to be limited, there was general agreement that it does occur and that plastic pipe should not be used in areas subject to organic chemical contamination. California does have a health department regulation prohibiting the use of plastic pipe under certain conditions considered to pose a possible adverse health effect. Plastic industry representatives stated that they are working on development of nonpermeable pipe materials. There are also new techniques being developed for improved installation, such as soil barriers, that might be helpful in preventing the permeation of plastic pipe.

Recommendations: There were general recommendations that the use of plastic pipe be limited in areas of existing or potential contamination; that industry or other bodies inform potential users regarding limitations of products relating to permeation, and that further investigation should be made into the development of nonpermeable pipe or other means of preventing pipe permeation.

INFORMATION TRANSFER AND COMMUNICATION

From its inception, this seminar represented a first meeting of many varied interests in which government, industry, plumbing experts, etc., took an active part in considering drinking water corrosion contamination problems and possible solutions. A common observation was that insufficient information transfer and dialogue had occurred in the past, particularly between the governmental water supply interests and the industrial and plumbing sectors.

Further, it was felt that insufficient time was available at this seminar and that the subjects covered were too broad to allow adequate time for meaningful discussion and resolution of issues.

<u>Recommendations</u>: There was a general recommendation for more study and deliberation in the substantive areas and for subsequent meetings planned with more focussed discussions and more time allowed. A further recommendation was made for more formal communication such as articles, papers, joint small discussions, etc., on plumbing materials and water quality, so that problems and solutions can be properly communicated among all the varied interests.

Part VI

Summation

SUMMATION—PLUMBING MATERIALS AND DRINKING WATER QUALITY

James M. Symons

Department of Civil Engineering
University of Houston, Central Campus
4800 Calhoun
Houston, Texas 77004

The Workshop was opened by Cotruvo, who introduced the subject and gave the themes of the program: to permit the airing of views and to provide directions for the future. Cotruvo suggested that the Workshop focus on three questions:

1. Do problems exist?
2. What technical solutions are available?
3. What are the roles of the various organizations?

I. DO PROBLEMS EXIST?

   A.   <u>Materials</u>

      1.   <u>Copper</u>

Anderson reported that copper piping has several good features: it is popular, functions well, has a good program of quality assurance, is based on the eddy current method, and is available during manufacture; also, the industry has developed techniques that remove the drawing chemicals. He noted, however, that copper pipe does give some problems; for instance, pitting can cause leakage, and corrosion will occur if utilities do not treat their water properly. This is particularly true in small utilities, where treatment is less sophisticated.

      2.   <u>Plastic</u>

Plastics were discussed by Mruk and Podoll, who noted that the favorable features of this material are its immunity to corrosion, the smooth surface of the piping material, the

lack of need for soldering joints, and the flexibility of
the pipe. Mruk pointed out some unfavorable factors,
however, such as problems with jointing materials, the
additives in the plastic pipe, its susceptibility to high
temperatures, its combustibility, and the reduction in
strength as the duration of loading increases. Podoll
studied chlorinated polyvinyl chloride (CPVC) pipe and found
that a few volatile organic compounds leached out at the 1
to 10-microgram/liter (ug/L) concentration. He also noted
that some jointing compounds were found during the leaching
test initially, but then were rinsed out. Organo-tin
initially leached at a concentration of 0.5 to 3 ug/L, but
the concentration fell to 0.1 ug/L after 3 weeks.

3.   Lead

None of the speakers or the panelists had many favorable
comments to make about lead pipe, but Karalekas reported on
severe lead problems in the New England area. He pointed
out that even 5.5 percent of running samples exceed the
regulated lead concentration. Karalekas discussed his
sampling protocol of a standing sample, a short-term running
sample, and a long-term running sample; several other
speakers and panelists mentioned the importance of
standardizing these sampling procedures as they do affect
the overall metal concentration in the resulting water
sample.

4.   Galvanized Pipe

Neff pointed out that new pipe dissolves more zinc into the
water than pipe that has been in place for a period of
approximately 2 years. Neff also pointed out that
chrome-plated brass valves, such as are found in the
household, produce metal concentrations in water; he
suggested that studies on piping systems should take into
account the contribution of metals from the valves and
faucets.

5.   Solders

Murrell presented data from a study on Long Island, where the
authors found that 67.3 percent of the piping systems had
solders containing from 55 to 65 percent lead content. They
found that this solder leaches metals on standing and that in
new homes, very high lead levels were initially found. They
also pointed out that slugs of water with high concentrations
of lead occurred periodically as water is drawn through a
plumbing system. As water that has been in contact with a
soldered joint comes out of the tap it will contain leached
metals.

B.   <u>Penetration Problems</u>

Pfau and Goodwin reported on the penetration of several solvents
through jointed and unjointed ductile iron, polyvinyl chloride
(PVC), and asbestos/cement (A/C) pipe. When the ground was
saturated with solvents only the solid ductile iron pipe resisted
penetration. All joints permitted the penetration of solvent
into the water and solid A/C pipe passed the solvent quite
easily. Solid PVC pipe also passed the solvent, but only after
32 days. Goodwin reported similar data when the pipe was
challenged with approximately 300 mg/L of solvent.

C.   <u>Importance of Water Chemistry and Environment</u>

Anderson noted that copper pipe would corrode if water velocities
are greater than 5 ft/sec, the water has a high concentration of
carbon dioxide, and the water is soft. Karalekas reported the
impact of soft water and low pH on lead solution. Neff indicated
that with galvanized piping, chlorine residual had an influence
on metal loss as did low pH, and commented that polyphosphates
may aggravate the metal loss situation. Murrell noted that low
pH and soft water caused increased problems from soldered joints.

II.   WHAT TECHNICAL SOLUTIONS ARE AVAILABLE?

A.   <u>Water Treatment</u>

Schock reviewed the impact of pH, carbonate concentrations,
orthophosphate, polyphosphate, and silicate on lead, copper, and
zinc. Schock pointed out that the solubility of lead actually
goes up as the carbonate concentration goes up, a find that might
not be intuitively obvious to the unsuspecting. He pointed out
that many people use the Langelier Index to calculate calcium
carbonate deposition, but that several factors in the Langelier
Index are based on rather old data; therefore, the actual pH of
saturation is frequently a few tenths of a pH unit different than
calculated by the traditional Langelier Index. He also pointed
out that polyphosphates are quite different from orthophosphate
and that in some cases polyphosphates may actually cause higher
metal content in water. Finally, he noted that silicates often
produce a favorable effect on galvanized pipe. Karalekas showed
several case histories in which the benefits of increasing pH had
a dramatic effect on the lead content in New England waters. The
importance of proper water treatment was also discussed in the
panels.

B. **Regulations**

Several alternative techniques (other than water treatment) for
controlling corrosion byproducts in water were suggested by
various speakers. For instance, Murrell recommended that lead
solder be banned and alternatives such as silver or tin-antimony
be used to avoid the dissolution of metal from the solder.
Boydston discussed altering plumbing codes as a way of
controlling the resultant water quality in household distribution
systems. He also pointed out that, in the absence of Federal
regulations, state and local ordinances could be passed to
control the use of certain plumbing materials. Wagner thoroughly
reviewed European practice, pointing out that many countries in
Europe had regulations restricting the use of materials that
would have a deleterious effect on water, for instance lead pipes
and lead solder. He also pointed out that plastic pipe material
was controlled by statute in many European countries. Another
approach to control would be Federal regulation by the U.S. EPA.
Currently many metals are regulated, either as primary or
secondary drinking water regulations and the corrosivity of water
should be measured. Vigorous corrective action through
enforcement after finding corrosive water would help reduce the
insult to the consumer from dissolution of metal. The importance
of education as a method of improving the current situation was
pointed out by the panels.

C. **Third-Party Review**

McClelland reported on the extensive review, inspection, and
testing protocol of the National Sanitation Foundation (NSF) for
plastic pipe. By investigating the manufacturing products that
went into pipe and the manufactured product itself, the NSF is
able to ensure that pipe containing their seal meets rather
rigorous standards. In any manufacturing product quality control
is important. Mruk, for example, told of subsequent problems
that resulted from a plastic pipe manufactured without a
necessary additive.

No matter which approach, or combination of approaches, is used
for controlling the deterioration of water quality during passage
through distribution piping and plumbing systems, inspection and
enforcement is essential to make this system work. For instance,
during the Workshop examples were given of plumbers who switch
from more desirable to less desirable solders when inspectors are
not looking; of water utilities that will not treat water to the
degree recommended by the consulting engineer; of pipe that may
be sold in an area where it is unsuitable (problems occurred some
years ago with asbestos/cement pipe in soft water areas; and of
installation practices not properly followed, for example, the
kinking of plastic pipe in an effort to have it bend beyond its
traditional flexibility.

III.   WHAT ARE THE ROLES OF THE VARIOUS ORGANIZATIONS?

All three parties represented at this Workshop have an important
interest in this subject. The manufacturers want their products to
enjoy a good reputation, both with the public and with the
regulators, so they can sell their product. The regulators want to
be able to offer users alternative piping materials and they want to
protect the health of the public. The independent service groups
such as the NSF provide an essential link between the other two
parties.

The Workshop was successful in generating a dialogue among the three
parties mentioned above. Only through cooperative effort, such as
the one demonstrated in this Workshop, can the ultimate goal of safe
water for the consumers be realized.

Other Noyes Publications

IDENTIFYING AND REDUCING LOSSES IN WATER DISTRIBUTION SYSTEMS

by

James W. Male, Richard R. Noss, I. Christina Moore
University of Massachusetts

This book provides guidelines for assessing those problems associated with the water supply purveyance system. Included are recommendations for determining water use, estimating water system losses, evaluating alternatives for water system inprovements, and ascertaining the benefits and costs involved.

In recent years, the cost of supplying water to cities and towns has increased substantially. These increasing costs, together with the threat of water shortages due to contamination of existing supplies and the difficulty of obtaining new supplies, have caused water system managers to give increasing attention to water conservation, leak detection and repair, maintenance programs and other internal "housekeeping" measures. However, before such measures are adopted, water system managers must determine if there is a problem with their water supply system, the magnitude and source(s) of the problem, and the efficacy of alternative solutions.

A condensed table of contents listing **chapter titles and selected subtitles** is given below.

1. INTRODUCTION, SUMMARY, AND CONCLUSIONS
General Terminology
Eliminating Water Losses
Record Keeping
Meter Accuracy
Leak Detection and Repair
Allocation of Funds

2. GUIDELINES FOR DETERMINING WATER SUPPLY AND CONSUMPTION
General Methodology and Terminology
Assessment of Water Use
  Water Supplied
  Water Consumed

3. INSPECTION AND MAINTENANCE PROGRAMS FOR 5/8 INCH WATER METERS
Assessing Meter Accuracy
  Data Collection and Analysis
  Working Meter Accuracy
  Meter Failure Rates
  Non-metered Water Estimates
  Meter Accuracy with Age
Optimum Periodic Meter Testing
  Data
  Optimum Period of Testing Model
  Overview of the Program ECONANAL
  Using the Program ECONANAL

4. DETERMINING LEAKAGE IN WATER DISTRIBUTION SYSTEMS
Leak Detection Methodology
  and Equipment
Survey of New England Water Utilities
  Survey Description
  System Characteristics
  Meter Management
  Unaccounted-for Water
Leak Detection Conclusions
  and Recommendations
Water Leakage Prediction Model

5. ECONOMICS OF LEAK DETECTION AND REPAIR: CASE STUDIES
Westchester Joint Water Works
Louisville Water Company

6. ALLOCATING FUNDS FOR LEAKY WATER DISTRIBUTION SYSTEMS: ECONOMIC AND INSTITUTIONAL CONSIDERATIONS
Allocation Procedures
  Maximize Water Saved
  Maximize Net Benefits
  Maximize Equity
  Constraints
Application of Procedures
  Results

REFERENCES

ISBN 0-8155-1050-0 (1985)   156 pages

CONTAMINANT REMOVAL
FROM PUBLIC WATER SYSTEMS

by

Daniel C. Houck et al

Pollution Technology Review No. 120

Based on studies by *D.H. Houck Associates; Rip G. Rice, Inc.; Wade Miller Associates; Purdue University and the City of Indianapolis, Indiana;* and *Environmental Science and Engineering, Inc.;* this book describes the rationale as well as methods for contaminant removal from public water systems. The book is for use by owners and operators, municipal managers and consulting engineers involved with the safe and efficient operation of public water systems. Its purpose is to assist personnel in understanding the importance of contaminant control and to explain design concepts, cost estimating techniques, and operational considerations associated with current technological approaches for maintaining such control.

The book is divided into four parts, each dealing with a distinct type of contaminant removal. Problem areas covered are microorganisms, nitrates, radionuclides, and turbidity.

The information provided is intended to explain:

1) why contaminant control is important,

2) theories of control,

3) process options for control,

4) design procedures for control,

5) process control methods,

6) operation and maintenance procedures,

7) cost estimation methods.

A condensed table of contents listing **part titles and selected subtitles** is given below.

I. MICROORGANISMS
Sources and Significance of
  Waterborne Disease
National Interim Primary Drinking
  Water Regulations (NIPDWR)
Maximum Contaminant Level (MCL)
  for Bacteriological Contaminants
Non-Treatment Alternatives
Treatment Alternatives
Disinfection with Chlorine,
  Chloramines, Chlorine Dioxide,
  Ozone, or Ultraviolet Radiation
Optimal System Design

Cost Estimating Procedures
Operation and Maintenance Practices
Manuals, Equipment, and Supplies
Monitoring
Preventive Maintenance
Emergency Procedures
Good Sanitary Practices
Safety Procedures
NIPDWR Compliance

II. NITRATES
When Nitrates Are a Problem
How Nitrates Get into Water Supplies
Treating Water Supplies for Nitrate
  Removal
Designing a Nitrate Removal System
Pilot Testing
Pretreatment Requirements
Construction Costs
Operation and Maintenance Costs

III. RADIONUCLIDES
Radionuclide Health Effects
Radionuclides in Drinking Water—
  Occurrence and Sources
Health Effects of Low Level
  Radioactivity in Drinking Water
Nontreatment Alternatives
Treating Water Supplies for Radium
  and Uranium Removal
Considerations in the Design of a
  Radionuclide Treatment System
Selection of a Radionuclide Treatment
  System
Pilot Studies for Evaluating
  Radionuclide Removal Processes
Characteristics of Waste Streams
  Generated
Construction Costs
Operation and Maintenance Costs

IV. TURBIDITY
When Turbidity Is a Problem
Definition and Causes of Turbidity
Nontreatment Alternatives
Treatment Alternatives for Turbidity
  Removal
Designing Turbidity Removal Systems
Cost Estimating Procedures and
  Funding Sources

REFERENCES

APPENDICES

ISBN 0-8155-1022-5 (1985)

524 pages